Air Distribution in Buildings

Essam E. Khalil

CRC Press
Taylor & Francis Group
Boca Raton London New York

CRC Press is an imprint of the
Taylor & Francis Group, an **informa** business

MECHANICAL and AEROSPACE ENGINEERING

Frank Kreith & Darrell W. Pepper
Series Editors

RECENTLY PUBLISHED TITLES

CRC Press
Taylor & Francis Group
6000 Broken Sound Parkway NW, Suite 300
Boca Raton, FL 33487-2742

First issued in paperback 2017

Version Date: 20130930

ISBN 13: 978-1-138-07664-8 (pbk)
ISBN 13: 978-1-4665-9463-0 (hbk)

Library of Congress Cataloging-in-Publication Data

Khalil, E. E.
    Air distribution in buildings / Essam E. Khalil.
        pages cm. -- (Mechanical and aerospace engineering series ; 54)
    Includes bibliographical references and index.
    ISBN 978-1-4665-9463-0 (hardback)
    1. Ventilation--Equipment and supplies--Design and construction. 2. Buildings--Ventilation. 3. Buildings--Air conditioning. I. Title.

TH7681.K53 2013
697.9'2--dc23   2013038089

Visit the Taylor & Francis Web site at
http://www.taylorandfrancis.com

and the CRC Press Web site at
http://www.crcpress.com

Contents

Preface

The present reference book gives comprehensive advice and guidance on the design, calculations, and efficient operation of air distribution in buildings of different natures and applications. The book focuses on buildings from a simple room to more complex configurations in archeological monuments to places of worship, such as churches and mosques. Special concern is given to hospitals, operating theaters, and sporting facilities, which are for use at various stages during the whole building life cycle. Air conditioning system providers have a duty of care to ensure that appropriate engineering governance arrangements are in place and are energy efficient. The author has spent the past 40 years working on the design of air conditioning and ventilation systems for monument facilities, as well as other electromechanical services. It is not the intention of this work to unnecessarily repeat any basic textbook material, industry-related practice, or government legislation. Where appropriate, these will be referenced. Managing the air conditioning systems in buildings is a vital tool in the safe and efficient operation of built environments. The core suite of six subject areas provides access to guidance that

- is more streamlined and accessible
- encapsulates the latest standards and best practice in air conditioning engineering
- provides a structured reference for built environment engineering

Building energy-efficient environments is a main concern and theme throughout the book. The reader is directed to energy-saving measures in air conditioning systems and different applications. Reference is always made to appropriate international standards and prevailing codes.

Acknowledgments

This work would not have been possible without the total cooperation of so many experienced individual colleagues, coupled with the backing of many well-known companies—what a wealth of professional knowledge and expertise! An acknowledgment such as this can only scratch the surface and cannot convey the very grateful thanks that are due to those people who have spent many hours of their valuable time putting together many pages of knowledge for the benefit of others. Those who seek advice and assistance in tackling their day-to-day problems, particularly young engineering students starting out on multidisciplinary careers, will be extremely grateful. Over the years I have put together many technical papers and essays, and it was with this in mind that I have structured the book. I have spent many hours trying to obtain information to assist me in my work, and hence the theory behind the structuring of this book.

I therefore take this opportunity to thank those colleagues in hospital engineering operations, consultants, and staff members for their patience and assistance in helping me to complete the book.

My family suffered considerably while this book was being written. My son Dr. Ahmed and daughters Prof. Heba, Eng. Rana, and Eng. Dalia have been more than tolerant, for which they deserve more than my thanks and appreciation.

About the Author

Essam E. Khalil, PhD, received his doctorate in mechanical engineering from the Imperial College of Science and Technology at the University of London in February 1977. He was then appointed as an assistant professor of mechanical engineering at Cairo University in Egypt and became a full professor in June 1988, working in the thermofluids research area. His research interests include turbulent combustion, heat transfer air conditioning, and energy-efficient building designs. At present, Dr. Khalil is a professor of energy and power plants and is currently the chair of the All Arab Air Conditioning Code Committee. He has supervised 30 PhD dissertations and more than 50 MS theses in the areas of air distribution in buildings, turbulent combustion, heat transfer, and sustainability. As a faculty member at the University of Cairo he received the Excellence in Research Award three times.

Dr. Khalil is an active fellow of the American Society of Mechanical Engineers (ASME), the American Society of Heating, Refrigerating, and Air Conditioning Engineers (ASHRAE), and the American Institute of Aeronautics and Astronautics (AIAA). He is regional director for ASHRAE in Africa and the Middle East subregion. He has served as principal investigator (PI) and co-PI in more than 10 research projects funded by the United States Agency for International Development (USAID), the Egyptian government, and private industrial entities. He is the convener of ISO TC205 WG2 on the design of energy-efficient built environments and ISO TC163 WG4 on the holistic approach. Currently, he is very active in energy and sustainability research and has been invited to give 20 plenary and keynote lectures in international conferences on energy and sustainability. Dr. Khalil is the author or coauthor of 11 textbooks, 3 book chapters, and more than 550 research papers in international journal and conference proceedings. He is an editorial board member for the *International Journal of Thermal and Environmental Engineering* and *Advances in Mechanical Engineering*. He received the ASME Westinghouse Gold Medal in 2009, AIAA Energy Systems Award in 2010, and ASME James Harry Potter Gold Award in 2012.

1

Air Distribution Systems

Air distribution systems usually comprise an air handling unit or fan coil unit that supplies air-conditioned air (treated air) to users through a network of conduits (usually rigid metallic ducts) feeding the air to terminal outlets such as grilles, diffusers, etc. Alternatively, air can be supplied from merely a supply fan that discharges air to ventilate a space; air is similarly conveyed through ducts to terminals.

1.1 Airflow Patterns

1.1.1 General

Central heating and cooling systems use an air distribution or duct system to circulate heated or cooled air to all the conditioned rooms in a building. Properly designed duct systems can maintain uniform temperatures throughout the building, efficiently and quietly. Air is generally issued from ceiling or wall-mounted jets. The air in the room is entrained and engulfed by the emerging jets; such air motion sets the necessary heat transfer among the various zones. As the air motion in the room is affected by the balance of inertia force, buoyancy, and viscous action, the airflow velocities have a determinant effect. Buoyancy has a pronounced effect on the flow pattern whenever the flow velocities are less than 0.25 m/s.

To design an optimum heating, ventilation, and air conditioning (HVAC) airside system that provides comfort and air quality in the air-conditioned spaces with efficient energy consumption is a great challenge. Air conditioning identifies the conditioning of air for maintaining specific conditions of temperature, humidity, and dust level inside an enclosed space. The conditions to be maintained are dictated by the need for which the conditioned space is intended and comfort of users. So, the air conditioning embraces more than cooling or heating. Comfort air conditioning is defined as the "process of treating air to control simultaneously its temperature, humidity, cleanliness, and distribution to meet the comfort requirements of the occupants of the conditioned space."[1] Air conditioning therefore includes the entire heat exchange operation, the regulation of velocity, thermal radiation, and quality of air, and the removal of foreign particles and vapors.[2] A successful HVAC design is one that is energy efficient, besides having

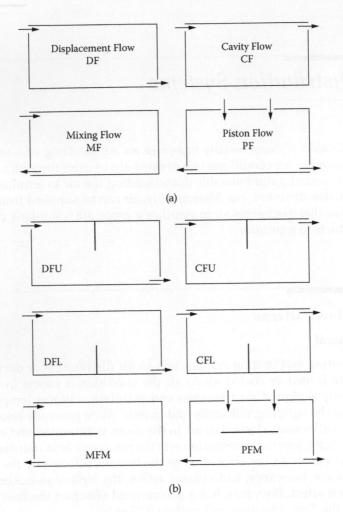

FIGURE 1.1
(a) Basic room supply and return/extract ports. (b) Room supply and return/extract ports.

all other comfort factors.[3–10] Various room air outlets designs are available. Figure 1.1 indicates the basic types:

1. Displacement flow
2. Cavity flow
3. Mixing flow
4. Piston flow

Displacement flow: Air is supplied from one side of the room (high wall) and pushes toward the extract or return air port at the opposite facing wall at a lower level, thus pushing and scavenging the air from across the room.

Cavity flow: Air is supplied from one side of the room (high wall) and pushes toward the extract or return air port at the opposite facing wall at a high level, thus allowing air to circulate in the formed cavity.

Mixing flow: Air is supplied from one side of the room (high wall) and pushes toward the opposite facing wall where it impinges and changes its direction toward the upstream supply wall where it is discharged at a lower level.

Piston flow: Air is supplied from the ceiling at one or more supply outlets and is discharged from extract ports on opposite walls near the floor. In such configuration the air is pushed from the ceiling to the floor.

Figure 1.2 shows the calculated flow pattern in terms of streamlines as reported earlier by the author.

The jet penetration expansion and diffusion in a room is strongly dependent on flow rate, turbulence intensity, supply outlet location, temperature gradients, and room thermal loads. The examples shown here in Figure 1.2 illustrate the streamline variation in the various cases. Piston flow (PF) configuration clearly indicated the sharp decay of the jet issued from the ceiling outlet even before it traveled 30% of its vertical descent. Figure 1.2b indicates the various turbulence characteristics in a displacement flow. Figure 1.2c shows the corresponding trajectories under heating and cooling conditions in a room 7 m long, 4 m wide, and 3 m high.

The corresponding isothermal line contours are shown in Figure 1.3; temperature can be readily increased from a supply value of 13°C at the ceiling to 18°C at 1 m above the finished floor in the 3 m high room.

It should be emphasized here that the selection, location, number, and types of supply outlets are very influential in determining the effectiveness of the air conditioning system in yielding comfort. It is important to achieve the following:

- No drafts or uncomfortable turbulence.
- Air velocities in the occupancy region should not exceed 0.25 m/s (at about 1.0 m above finished floor).
- No pockets of either a cold or hot region, i.e., homogeneous temperature distribution.

1.1.2 Practical Configurations

1.1.2.1 Grilles and Diffusers

These devices are usually located at the end of the air distribution system, at the far downstream end for supply air grilles and diffusers, as shown in

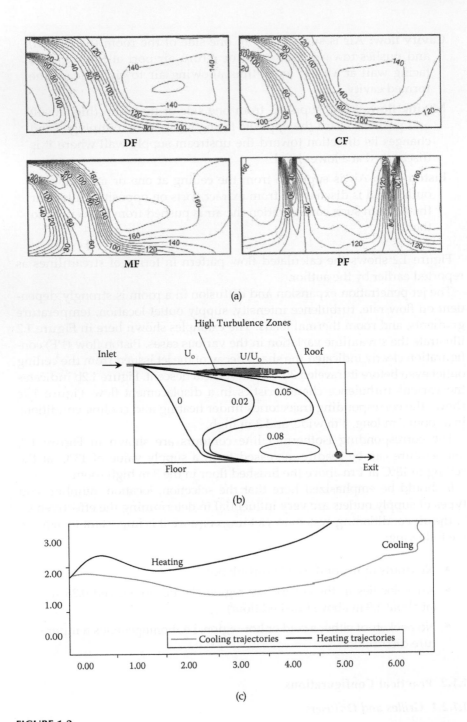

FIGURE 1.2
(a) Airflow pattern in a room with various air inlet and exit locations. (b) Airflow pattern in a room with displacement flow configurations. (c) Air trajectory in a room.

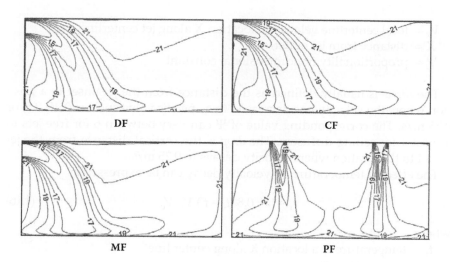

FIGURE 1.3
Air temperature distribution in a room with various air inlet and exit locations, in degrees °C.

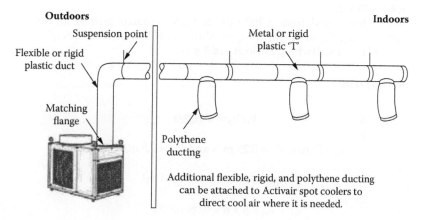

FIGURE 1.4
A typical duct connection comprising a rigid horizontal duct with flexible polythene duct branches.

Figure 1.4. The supply grilles and diffusers can be of different shapes and designs to achieve the goals shown in Figures 1.2 and 1.3.

The isothermal free jet flow behavior can be generally characterized by the equation

$$V_x/V_0 = \Psi \, A_0^{1/2}/x \tag{1.1a}$$

where
$V_0 =$ initial jet velocity, m/s
$A_0 =$ area corresponding to initial velocity, m²

V_x = local centerline velocity at location X along jet centerline, m/s
X = distance from jet outlet, m
Ψ = proportionality nondimensional constant

The *throw* is usually defined as the distance from the jet onset to a position where the velocity has decreased to a specific value such as 0.25, 0.5, or 0.75 m/s. The corresponding value of Ψ can vary between 6 for free jets to almost 1 for ceiling diffusers. The *drop* is the vertical distance from the jet outlet to the location where velocity decays to 0.25 m/s.

The effect of temperature on velocity decay can be expressed as

$$t_x - t_r = 0.8\,(t_0 - t_r)\,V_x/V_0 \qquad\qquad (1.1b)$$

where
t_x = temperature at a location X along center line
t_r = room temperature
t_0 = outlet temperature

EXAMPLE 1.1
Air is discharged from a 100 mm diameter circular pipe to an open space. Using Equation 1.1, compare throw at terminal velocities of 0.25 and 0.5 m/s for discharge velocity of 5 m/s.

SOLUTION:

$$V_x/V_0 = \Psi\,A_0^{\frac{1}{2}}/x$$

$$V_0 = 5 \text{ m/s}, V_{x1} = 0.25 \text{ m/s and } V_{x2} = 0.5 \text{ m/s}, \Psi = 6$$

$$A_0 = \pi/4\,(0.1)^2 = 0.007854 \text{ m}^2$$

$$V_{x1} = 5 \times 6 \times 0.08862/X_1$$

$$X_1 = 10.63 \text{ m, at which } V_{x1} = 0.25 \text{ m/s}$$

EXAMPLE 1.2
Air is discharged at 45°C into a room whose temperature is 20°C at velocity 4 m/s. Calculate the temperature at a distance 2 m and 10 m.

SOLUTION:

$$V_x/V_0 = \Psi\,A_0^{\frac{1}{2}}/x \text{ and } t_x - t_r = 0.8\,(t_0 - t_r)\,V_x/V_0$$

$$t_r = 20°C, t_0 = 45°C$$

$$x = t_r + 0.8\,(t_0 - t_r)\,V_x/V_0$$

$$t_x = t_r + 0.8 \, (t_0 - t_r) \, \Psi \, A_0^{\frac{1}{2}}/x$$

$$= 20 + 0.8(45 - 20) \, 6 \times 0.08862/2$$

$$= 25.317°C$$

At 10 m, $t_{10} = 21.06°C$.

EXAMPLE 1.3

Illustrate the jet development in a Piston type flow in a room. Use the sketches shown below:

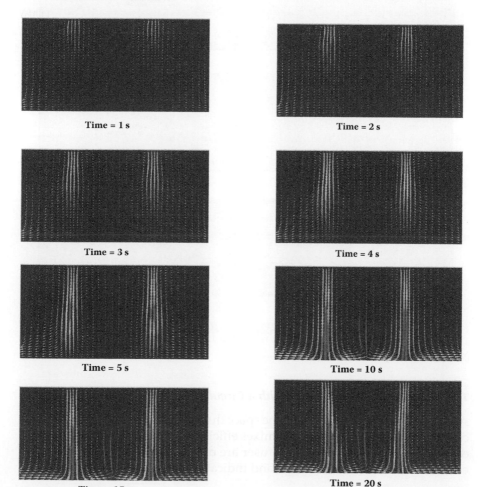

Time = 1 s   Time = 2 s

Time = 3 s   Time = 4 s

Time = 5 s   Time = 10 s

Time = 15 s   Time = 20 s

-0.86 -0.75 -0.64 -0.53 -0.42 -0.31 -0.2 -0.09 0.02 0.13   W m/s

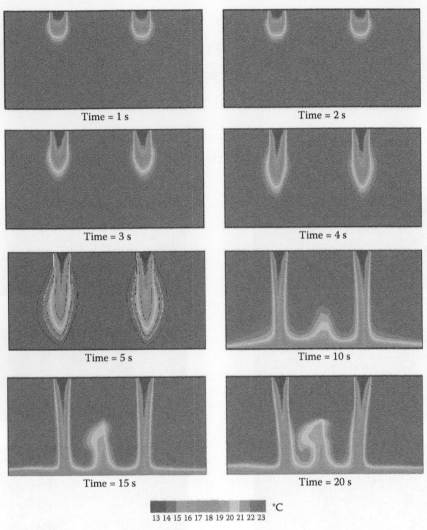

Time = 1 s

Time = 2 s

Time = 3 s

Time = 4 s

Time = 5 s

Time = 10 s

Time = 15 s

Time = 20 s

13 14 15 16 17 18 19 20 21 22 23   °C

(See color insert.)

1.1.2.2 Square Ceiling Diffuser with a Circular Duct Connection

Air is supplied horizontally into the space through slots in the diffuser's coni-
cal front panel. Induced room air mixes efficiently with supply air outside the
device. The fixed vanes of the diffuser are designed to maintain a "coanda"
effect. Figure 1.5 shows this type and indicates the expected flow pattern.

1.1.2.3 Wall-Mounted Supply Grille

Air is supplied into the space through the vanes, and mixes with room air
in front of the grille. Air is supplied straight or at an angle of 15° upward;

Hr [m]	Lr/Wr [m]	Hs [m]	ΔT [°C]	qv [l/s]	a	....			
						[m/s]	[m/s]	[m/s]	[m/s]
3.0	8.0	3.0	−8	65.0	0.0	0.00	0.20	0.40	0.60

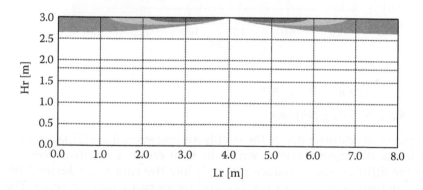

FIGURE 1.5
Flow pattern and design of ceiling diffuser.

the desired flow pattern is adjusted by changing the angle of the rear vanes. In wall installations, the recommended distance from the ceiling is 200 mm, when the supply air is directed to the ceiling surface. The grille can also be for exhaust air (Figure 1.6).

1.1.2.4 Ceiling Architectural Diffuser

Air is supplied horizontally into the space through the diffuser slots (1, …, 4). Supply air mixes efficiently with the room air outside the diffuser. The fixed vanes of the diffuser are designed to maintain a coanda effect, even with variable air volume operation (Figure 1.7).

Hr [m]	Lr/Wr [m]	Hs [m]	ΔT [°C]	qv [l/s]	alfa °	a	[m/s]	[m/s]	[m/s]	[m/s]
6.0	16.0	4.0	−8	550.0	15.0	0.0	0.64	0.20	0.40	0.60

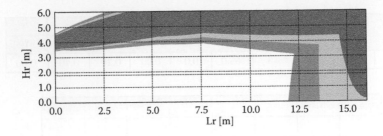

FIGURE 1.6
Wall-mounted supply grille flow pattern.

1.1.2.5 Square Ceiling Diffuser

The cones of the diffuser divide the supply air into several jets. This arrangement forms divergent sections, which in turn create a negative pressure under the diffuser and so induce room air into the supply air device. This internal induction reduces air velocity and temperature into the space. The process is repeated outside of the diffuser between the supply and mixed room air, further reducing velocity and the temperature difference between supply and room air (Figure 1.8).

1.1.2.6 High Ceiling Diffusers

The supply air jet can either be directed radially across the underside of a false ceiling or, by changing the position of the supply cones in relation to the fixed outer frame, be blown downwards (Figures 1.9 and 1.10).

1.1.2.7 Fixed Swirl Diffusers for Ceiling Installations

Air is supplied into the space horizontally through the profiled blades of the device. The swirl effect induces supply air to mix with room air outside

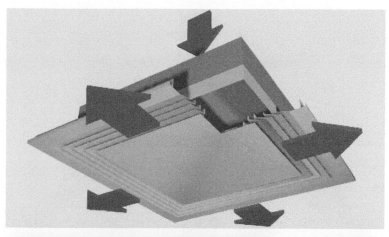

Hr	Lr/Wr	Hs	ΔT	qv	a				
[m]	[m]	[m]	[°C]	[l/s]		[m/s]	[m/s]	[m/s]	[m/s]
3.0	8.0	3.0	−8	110.0	0.0	0.27	0.20	0.40	0.60

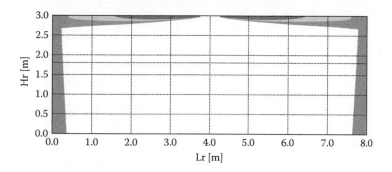

FIGURE 1.7
(**See color insert.**) Ceiling architectural diffuser.

the device, rapidly reducing the velocity of the supply air jet and efficiently balancing the temperature in the space (Figure 1.11). This diffuser is recommended in spaces where high heat loads require efficient air circulation.

1.2 Airflow Efficiency in Rooms

The airflow regimes in the room can be viewed as carriers of energy, and the following efficiencies can be used. The thermal performance of airflow can be evaluated using two different indexes. These indexes indicate the heat removal efficiencies and the energy efficiency. The energy efficiency evaluates

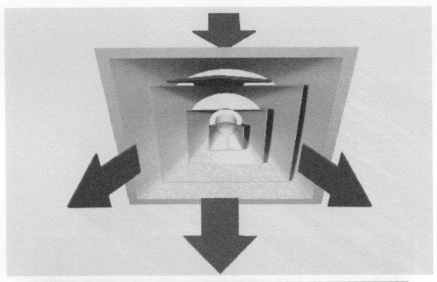

Hr	Lr/Wr	Hs	ΔT	qv	a				
[m]	[m]	[m]	[°C]	[l/s]		[m/s]	[m/s]	[m/s]	[m/s]
3.0	8.0	3.0	−8	55.0	0.0	0.00	0.20	0.40	0.60

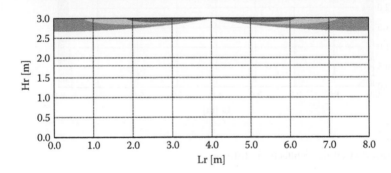

FIGURE 1.8
Square ceiling diffuser flow pattern and design.

the performance of the airside design for air utilization in a confined room configuration. This index is proposed to evaluate the airside design, and the effectiveness of the extraction port(s) location.

Heat removal efficiency (E_h) can be defined as

$$E_h = (T_e - T_i)/(T_m - T_i) \qquad (1.2)$$

Energy efficiency (E_e) is defined as

$$E_e = (T_e - T_i)/(T_t - T_i) \qquad (1.3)$$

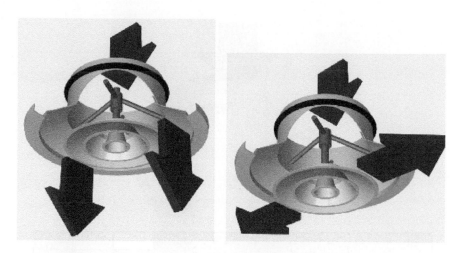

FIGURE 1.9
High ceiling diffuser design.

Hr	Lr/Wr	Hs	ΔT	qv	a				
[m]	[m]	[m]	[°C]	[l/s]		[m/s]	[m/s]	[m/s]	[m/s]
20.0	30.0	20.0	−8	35.0	0.0	0.00	0.20	0.40	0.60

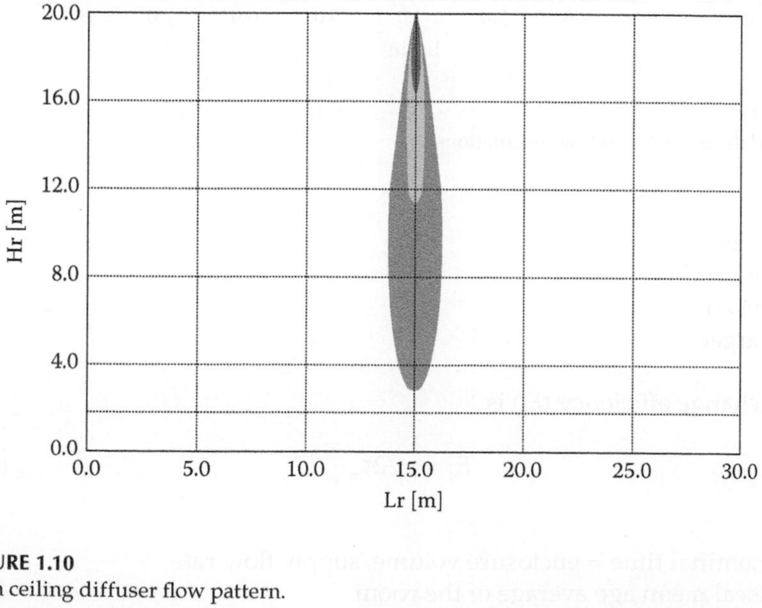

FIGURE 1.10
High ceiling diffuser flow pattern.

Hr	Lr/Wr	Hs	ΔT	qv	a				
[m]	[m]	[m]	[°C]	[l/s]		[m/s]	[m/s]	[m/s]	[m/s]
3.0	8.0	3.0	−8	30.0	0.0	0.00	0.20	0.40	0.60

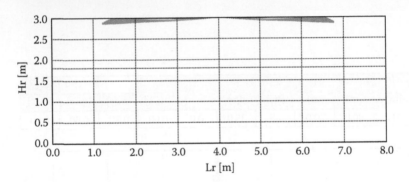

FIGURE 1.11
Fixed swirl diffusers for ceiling installations.

where:

 i = inlet
 e = extraction
 m = mean
 t = target

Air exchange efficiency (E_a) is

$$E_a = \tau_n / 2\tau_m \qquad (1.4)$$

where:

 τ_n = nominal time = enclosure volume/supply flow rate
 τ_m = local mean age average of the room

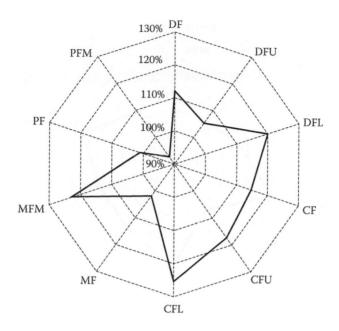

FIGURE 1.12
Heat removal efficiency E_h.

The airflow distribution efficiency can be generally assessed using the air exchange efficiency as shown in Figures 1.12 to 1.15 after Kameel and Khalil,[11] with numerical procedures outlined in references 12 to 16.

1.3 Noise Criteria

Noise produced by air grilles and supply outlets contributes to discomfort in the room and should be avoided at all times. To identify the spectrum content on noise, the use of noise criteria (NC) is proposed. Generally levels below NC 30 are considered to be quite; noisy conditions refer to NC greater than 50 to 55. Table 1.1 indicates the ASHRAE recommended noise criteria for various applications.

Generally, design goals for air conditioning system sound control for indoor areas will be in accordance with the ASHRAE system handbook[17] as summarized below; the design should incorporate provisions for sound control in order to achieve an appropriate sound level for all activities and people involved.

The above values are for unoccupied spaces with all air conditioning. Outdoor noise levels should not exceed the local environmental protection laws less than 60 db (A) at 3 m.

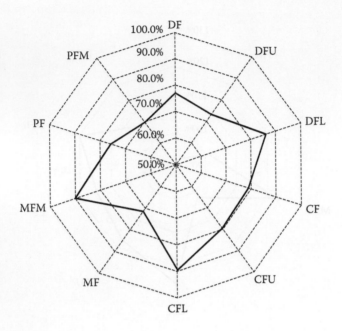

FIGURE 1.13
Energy efficiency E_e.

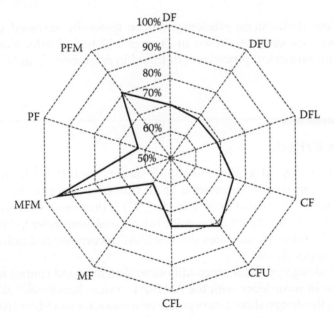

FIGURE 1.14
Air exchange efficiency E_a.

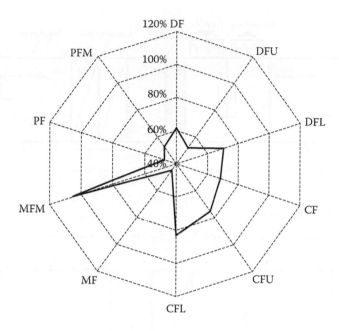

FIGURE 1.15
Gross efficiency.

TABLE 1.1

Recommended Noise Criteria (ASHRAE)

Type of Area	RC or NC Criteria Range
Dormitories and apartments	30–35
Individual rooms or suites	30–35
Meeting/banquet rooms	30–35
Halls, corridors, lobbies, and services	40–45
Executive offices and clinics	25–30
Conference rooms	25–30
Private area	30–35
Open offices and secretaries	35–40
Administrative assistant	35–40
Lobbies and circulation	35–40
Computer/business machines	40–45
Public circulation	40–45
Banking areas	30–35
Dining rooms	35–40
Athletic activity rooms	40–45

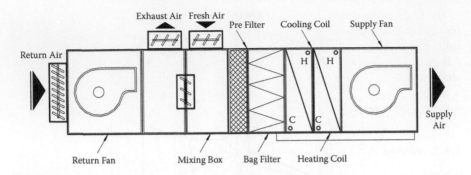

FIGURE 1.16
Circulating air handling units.

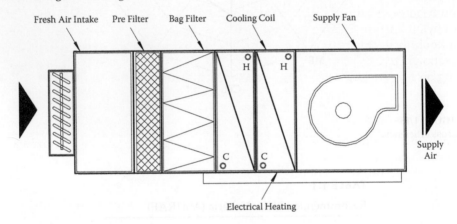

FIGURE 1.17
One hundred percent fresh air handling units.

1.4 Air Handling Units

Air handling units are devices that treat the air and are generally con-structed of modular patterns of added sections that yield the main functions of supplying, filtering, cooling, humidification/dehumidification, heating, reheating, extra filtering of the air, etc. These should provide air that yields thermal comfort in occupied zones.[18–22] The air handling unit would typically comprise the following, as shown in Figures 1.16 and 1.17.

1.4.1 Casing

Unit casing shall be a 23 mm double-skinned sandwich construction with a 0.6 mm inner galvanized skin and a 1.0 mm outer aluminum finish. The cavity shall be pressure injected with foam insulation, density 40 kg/m³, and k factor of 0.02 W/m°C.

The main aluminum framework shall incorporate and insulate the twin box section to prevent thermal bridging and condensation on the outside of the unit. The base frame shall be constructed from 4 mm thick aluminum alloy. All panels shall be detachable or hinged, depending on unit size. Hinges shall be manufactured from die-cast aluminum with stainless steel pivots.

Multisection units are usually suitable for on-site assembly and incorporate the necessary gaskets. All fixings gaskets shall be concealed.

1.4.2 Mixing Box

If required, the mixing box shall be complete with automatic opposed blade dampers. Blades shall be made of double-skinned airfoil aluminum extruded sections with integral gaskets and assembled within a rigid extruded aluminum frame. All linkages and supporting spindles shall be made of aluminum or nylon, turning in Teflon bushes. Manual dampers shall be provided within a knob for locking the damper blades in position. Damper frames shall be sectionalized to minimize blade wrapping. Dampers shall be sized depending on the air volumes, with maximum face velocities not exceeding 5 m/s.

1.4.3 Filter Sections

Each filter section shall be easily accessible and designed for easy withdrawal and renewal of filter cells. The filter framework should be fully sealed and constructed from aluminum alloy to specifically suit the filter types detailed in clause B.8. Filter face velocities shall not exceed manufacturer's recommendations. Filter types and efficiencies shall conform to EUROVENT 4-5 as specified. Manometers, supplied with the air handling unit, shall be fitted across all filter banks.

1.4.4 Coil Section Two-Pipe System

Coils shall be constructed from copper tubes, mechanically bonded to aluminum fins, and assembled within a heavy-gauge aluminum framework. Headers shall be copper. Coils shall be provided with air vent and drain plugs and leak tested at not less than 28 bar. Coil assemblies shall be supported on bearings for withdrawal. The insulated drain pan shall be full width and constructed from 1.5 mm thick aluminum sheeting with an oversize drain outlet. Coil face velocities shall not exceed 2.5 m/s with maximum water-side pressure drop of 30 kPa for both heating and cooling coils. Coil performances shall be based on chilled water at 6°C flow and 12°C return.

1.4.5 Electric Heater Section

This section should have at least three steps at a suitable rating, complete with controls, protection, and safety devices. An electric repeater section may be used, when necessary, to control the relative humidity, complete with controls, protection, and safety devices.

1.4.6 Fan Section

Fans shall be centrifugal, with forward-curved or backward-curved airfoil blades. Fan casings and wheels shall be made from aluminum alloy or mild steel sheets dependent on size and type. All backward-curved fans shall be provided with aluminum inlet cones. Fan shafts shall be preground stainless steel and supported in self-aligning plumber block grease-lubricated bearings. Fan wheels and pulleys shall be individually tested and precision balanced statically and dynamically and keyed to the shaft. The manufacturer must provide for curves showing the performance that will be achieved when tested. Similarly, data must be submitted showing the sound power level for all eight octave bands that will be achieved when tested.

The fan curves shall clearly show the duty point and absorbed power to ensure stable operating conditions and a minimum margin of 20% between absorbed and motor power. Motors shall be mounted on slide rails for easy belt tensioning, and be totally enclosed and fan cooled with F class insulation. Twin-speed motors shall be made available where specified. Motors shall drive heavy duty V-belts. Both fan and motor assemblies shall be mounted on a deep section aluminum alloy base frame. Isolation shall be provided from the unit casing by spring antivibration (AV) mounts (96% isolation efficiency) and a flame-retardant, waterproof neoprene-impregnated flexible connection on the fan discharge. Additional flexible connectors shall be provided on the unit outlet. Units shall be suitable on 380 V three-phase 50 Hz electrical supply.

References

1. ASHRAE. 2009. *Fundamentals*. ASHRAE, Atlanta, GA.
2. Sandberg, M. 1981. What is ventilation efficiency? *Building and Environment* 16, 123–135.
3. Khalil, E. E. 2010, January. *Thermal comfort and air quality in sustainable climate controlled healthcare applications*. AIAA-2010-0802. Orlando, FL.
4. Kameel, R. 2002. Computer aided design of flow regimes in air-conditioned operating theatres. Ph.D. thesis, Cairo University.
5. Khalil, E. E. 2002. *Energy efficiency in air-conditioned operating theatres*. IECEC 2002, paper 20050. ASME.
6. Khalil, E. E. 1978. Numerical procedures as a tool to engineering design. *Informatica* 78.
7. Kameel, R., and Khalil, E. E. 2002. Predictions of flow, turbulence, heat transfer and humidity patterns in operating theatres. ROOMVENT 2002.

8. Sekhar, S. C., Wai, T. K., Cheong, D., and Hien, W. N. 2001. Indoor air quality, ventilation and energy studies in hot and humid climates. Clima 2000, Napoli.
9. Medhat, A. A. 1999. Air flow patterns in air conditioned rooms. Ph.D. Thesis, Cairo University.
10. Khalil, E. E. 1999. Fluid flow regimes interactions in air conditioned spaces. *Proceedings of the 3rd Jordanian Mechanical Engineering Conference*, Amman, May 1999, p. 79.
11. Kameel, R., and Khalil, E. E. 2003. *Simulation of flow, heat transfer and relative humidity characteristics in air-conditioned surgical operating theatres.* AIAA-2003-0858.
12. Spalding, D. B., and Patankar, S. V. 1974. A calculation procedure for heat, mass and momentum transfer in three dimensional parabolic flows. *International Journal of Heat and Mass Transfer* 15, 1787.
13. Launder, B. E., and Spalding D. B. 1974. The numerical computation of turbulent flows. *Computer Methods in Applied Mechanics* 3(2), 269–275.
14. Kameel, R., and Khalil, E. E. 2003. *Air flow regimes in operating theatres for energy efficient performance.* AIAA-2003-0686.
15. Huzayyin, O. A. S. 2005. Flow regimes and thermal patterns in air conditioned operating theatres. M.Sc. thesis, Cairo University.
16. Nielsen, P. V. 1989. *Numerical prediction of air distribution in rooms. Building systems: Room air and air contaminant distribution.* ASHRAE.
17. ASHRAE. 2011. *Systems.* ASHRAE, Atlanta, GA.
18. Tanabe, S., Kimura, K., and Hara, T. 1987. Thermal comfort requirement during the summer season in Japan. *ASHRAE Transactions* 93(1), 564–577.
19. Nevins, R., Gonzalez, R. R., Nishi, Y., and Gagge, A. P. 1975. Effect of changes in ambient temperature and level of humidity on comfort and thermal sensations. *ASHRAE Transactions* 81(2).
20. Khalil, E. E. 2000. Computer aided design for comfort in healthy air conditioned spaces. *Healthy Buildings 2000*, Finland, 2, 461–466.
21. ASHRAE Standard 62-2010. 2010. *Ventilation for acceptable indoor air quality.*
22. ASHRAE Standard 55-2010. 2010. *Thermal comfort.*

2

Mathematical Modeling Technique

The present chapter has been prepared to describe and discuss the problems associated with the formulation of the equations appropriate to the airflow distribution indoors, in terms of the governing conservation equations and turbulence modeling assumptions and the computational procedure required to solve the equations with boundary and inlet conditions. The elliptic partial differential equations, which govern the transport of mass, momentum, and energy, in addition to their derivations, are presented later; they are restricted here to steady and three-dimensional configurations where recirculation may occur, such as in the airflow in the air-conditioned applications. These time-averaged equations contain, for turbulent flows, second-order correlations of fluctuating properties, and models to determine these correlations are necessary to make the equations soluble. The governing differential equations, expressed in finite difference form, are solved numerically by an iterative procedure, which is described in detail later. The turbulence models embodied in the numerical computational techniques to represent these unknown correlations are discussed and the appropriate modifications suggested. The selected turbulence model, in the form of a set of steady partial differential equations, allows the predictions of the aerodynamics properties of the flow. The various assumptions, limitations, and required convergence criteria to satisfy the conservation equations are also discussed later.

2.1 Governing Equations

2.1.1 Conservation of Mass: The Continuity Equation

2.1.1.1 Derivation of the Continuity Equation

A system is defined as a collection of unchanging contents, so the conservation of mass principle for a system is simply stated as time rate of change of the system mass equals zero.[1-4]

$$\frac{DM_{sys}}{Dt} = 0 \qquad (2.1)$$

where the system mass, M_{sys}, is more generally expressed as

$$M_{sys} = \int\limits_{sys} \rho \, d\mathcal{V}$$ (2.2)

where the integration is over the volume of the system. In words, Equation 2.2 states that the system mass is equal to the sum of all the density-volume element products for the contents of the system.

For a system and a fixed, nondeforming control volume that are coincident at an instant of time, the Reynolds transport theorem allows us to state that

$$\frac{D}{Dt} \int\limits_{sys} \rho \, d\mathcal{V} = \frac{\partial}{\partial t} \int\limits_{cv} \rho \, d\mathcal{V} + \int\limits_{cs} \rho V \cdot \hat{n} \, dA$$ (2.3)

Equation 2.3 can express the time rate of change of the system mass as the sum of two control volume quantities, the time rate of change of the mass of the contents of the control volume, and the net rate of mass flow through the control surface. It is convenient to use the control volume approach for fluid flow problems, with the control volume representation of the conservation of mass written as

$$\frac{\partial}{\partial t} \int\limits_{cv} \rho \, d\mathcal{V} + \int\limits_{cs} \rho V \cdot \hat{n} \, dA = 0$$ (2.4)

where the equation (commonly called the continuity equation) can be applied to a finite control volume (cv), which is bounded by a control surface (cs). The first integral of the left side of Equation 2.4 represents the rate at which the mass within the control volume is increasing, and the second integral represents the net rate at which mass is flowing out through the control surface (rate of mass outflow – rate of mass inflow). To obtain the differential form of the continuity equation, Equation 2.4 is applied to an infinitesimal control volume.

2.1.1.2 Differential Form of the Continuity Equation[1-4]

An orthogonal cubical control volume is considered as shown in Figure 2.1. At the center of the fluid density element is ρ and the velocity has components U, V, and W. Since the element is small the volume integral in Equation 2.4 can expressed as

$$\frac{\partial}{\partial t} \int\limits_{cv} \rho \, d\mathcal{V} \approx \frac{\partial \rho}{\partial t} \delta x \, \delta y \, \delta z$$ (2.5)

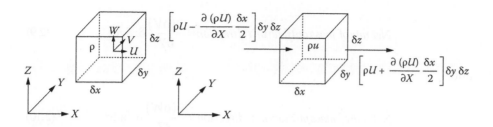

FIGURE 2.1
A differential element for the development of conservation of mass equation.

The rate of mass flow through the surfaces of the element can be obtained by considering the flow in each of the coordinate directions separately. If the term ρU represents the X component of the mass rate flow per unit area at the center of the element, then on the right face

$$\rho U\big|_{x+(\delta x/2)} = \rho U + \frac{\partial(\rho U)}{\partial X}\frac{\delta x}{2} \tag{2.6}$$

And on the left face,

$$\rho U\big|_{x-(\delta x/2)} = \rho U - \frac{\partial(\rho U)}{\partial X}\frac{\delta x}{2} \tag{2.7}$$

That was based on Taylor series expansion of ρU and neglecting higher-order terms. When the right-hand sides of Equations 2.6 and 2.7 are multiplied by the area $\delta y\,\delta z$, the rate at which mass crossing the right and left sides of the element is obtained is as shown in Figure 2.1. When these two expressions are combined, the net rate of mass flowing from the element through the two surfaces can be expressed as

$$Net\ rate\ of\ mass\ outflow\ in\ X\ direction = \left[\rho U + \frac{\partial(\rho U)}{\partial X}\frac{\delta x}{2}\right]\delta y\delta z$$

$$-\left[\rho U - \frac{\partial(\rho U)}{\partial X}\frac{\delta x}{2}\right]\delta y\delta z$$

$$= \frac{\partial(\rho U)}{\partial X}\delta x\delta y\delta z \tag{2.8}$$

For simplicity, only the flow in the X direction has been considered in Figure 2.1, but in general, there will also be flow-governing equations in the Y and Z directions. An analysis similar to the one used for flow in the X direction shows that

$$\text{\textit{Net rate of mass outflow in Y direction}} = \frac{\partial(\rho V)}{\partial Y} \delta x \delta y \delta z \qquad (2.9)$$

and

$$\text{\textit{Net rate of mass outflow in Z direction}} = \frac{\partial(\rho W)}{\partial Z} \delta x \delta y \delta z \qquad (2.10)$$

Thus,

$$\text{\textit{Net rate of mass outflow}} = \left[\frac{\partial(\rho U)}{\partial X} + \frac{\partial(\rho V)}{\partial Y} + \frac{\partial(\rho W)}{\partial Z} \right] \delta x \delta y \delta z \qquad (2.11)$$

From Equations 2.4, 2.5, and 2.11 it now follows that the differential equation for conservation of mass is

$$\frac{\partial \rho}{\partial t} + \frac{\partial(\rho U)}{\partial X} + \frac{\partial(\rho V)}{\partial Y} + \frac{\partial(\rho W)}{\partial Z} = 0 \qquad (2.12a)$$

As previously mentioned, this equation is also commonly referred to as the continuity equation. The continuity equation is one of the fundamental equations of fluid mechanics, and as expressed in Equation 2.12a, is valid for steady or unsteady flow. For steady flow, the equation will be

$$\frac{\partial(\rho U)}{\partial X} + \frac{\partial(\rho V)}{\partial Y} + \frac{\partial(\rho W)}{\partial Z} = 0 \qquad (2.12b)$$

2.1.2 Newton's Second Law: The Linear Momentum Equation[1-4]

2.1.2.1 Derivation of the Linear Momentum Equation

Newton's second law of motion for a system is the time rate of change of the linear momentum of the system equals to sum of external forces acting on the system.

$$\frac{D}{Dt} \int_{sys} V \rho dV = \sum F_{sys} \qquad (2.13)$$

Any reference or coordinate system for which this statement is true is called inertial. A fixed coordinate system is inertial. A coordinate system that moves in a straight line with constant velocity and is thus without acceleration is also inertial. When a control volume is coincident with a system at an instant of time, the forces acting on the system and the forces acting on the contents of the coincident control volume are instantaneously identical.[1,2]

Furthermore, for a system and the contents of a coincident control volume that are fixed and nondeforming, the Reynolds transport theorem allows concluding that

$$\frac{D}{Dt} \int_{sys} V\rho \, dV = \frac{\partial}{\partial t} \int_{cv} V\rho \, dV + \int_{cs} V\rho V \cdot \hat{n} \, dA \qquad (2.14)$$

Equation 2.14 states that the time rate of change of system linear momentum is expressed as the sum of two control volume quantities, the time rate of change of the linear momentum of the contents of the control volume, and the net rate of linear momentum flow through the control surface. As particles of mass move into or out of a control volume through the control surface, they carry linear momentum in or out. Thus, linear momentum flow should seem to be no more usual than mass flow.

For a control volume that is fixed (inertial) and nondeforming, an appropriate mathematical statement of Newton's second law of motion is

$$\frac{\partial}{\partial t} \int_{cv} V\rho \, dV + \int_{cs} V\rho V \cdot \hat{n} \, dA = \sum F_{contents\,of\,the\,control\,volume} \qquad (2.15)$$

$$F = \frac{DP}{Dt}\bigg|_{sys} \qquad (2.16)$$

F is result force acting on a fluid mass, *P* is linear momentum

Equation 2.15 could be applied to a finite control volume to solve a variety of flow problems. To obtain the differential form of the linear momentum equation (2.16) to a differential system, consider mass, δ*m*, or apply Equation 2.15 to an infinitesimal control volume, δ*V*, which initially bounds the mass δ*m*. It is probably simpler to use the system approach since application of Equation 2.16 to the differential mass, δ*m*, yields

$$\delta F = \frac{D(V\delta m)}{Dt} \qquad (2.17a)$$

where δ*F* is the resultant force acting on δ*m*. Using this system approach, δ*m* can be treated as a constant so that

$$\delta F = \delta m \frac{DV}{Dt} = \delta m \, a \qquad (2.17b)$$

2.1.2.2 *Description of Forces Acting on Differential Element*

In general, two types of forces need to be considered: surface forces, which act on the surface of the differential element, and body forces, which are distributed throughout the element. Body force is the weight of the element, which can be expressed as

$$\delta F_b = \delta m \ g \tag{2.18a}$$

g is the vector of the acceleration of gravity

$$\delta F_{bx} = \delta m \ g_x \tag{2.18b}$$

$$\delta F_{by} = \delta m \ g_y \tag{2.18c}$$

$$\delta F_{bz} = \delta m \ g_z \tag{2.18d}$$

g_x, g_y, and g_z are the components of the acceleration of gravity vector in X, Y, and Z directions.

Surface forces act on the element as a result of its interaction with its surroundings. At any arbitrary location within a fluid mass, the force acting on a small area, δA, which lies in an arbitrary surface, can be represented by δF_s. In general, δF_s will be inclined with respect to the surface. The force δF_s can be resolved into three components, δF_n, δF_1, and δF_2, where δF_n is normal to the area, δA, and δF_1 and δF_2 are parallel to the area and orthogonal to each other. The normal stress, σn, is defined as

$$\sigma_n = \lim_{\delta A \to 0} \frac{\delta F_n}{\delta A} \tag{2.19a}$$

And the shearing stresses are defined as

$$\tau_1 = \lim_{\delta A \to 0} \frac{\delta F_1}{\delta A} \tag{2.19b}$$

$$\tau_2 = \lim_{\delta A \to 0} \frac{\delta F_2}{\delta A} \tag{2.19c}$$

We will use σ for normal stress and τ for shearing stresses. A normal stress and two shearing stresses can thus characterize the intensity of the force per unit area at a point in a body. One can express the surface forces acting on a small cubical element of fluid in terms of the stresses acting on the

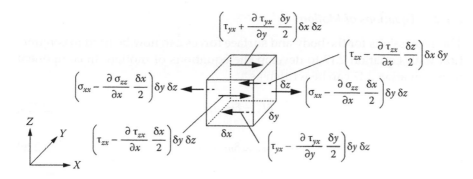

FIGURE 2.2
Surface forces in the X direction acting on a fluid element.

faces of the element as shown in Figure 2.2. It is expected that, in general, the stresses will vary from point to point within the flow field. Thus, we will express the stresses on the various faces in terms of the corresponding stresses at the center of the element of Figure 2.2 and their gradients in the coordinate directions. For simplicity, only the forces in the X direction are shown. Note that the stresses must be multiplied by the area on which they act to obtain the force. Summing all these forces in the X direction yields the resultant surface force in the X direction.

In a similar manner the resultant surface forces in the Y and Z directions can be obtained and expressed as

$$\delta F_{sx} = \left(\frac{\partial \sigma_{xx}}{\partial x} + \frac{\partial \tau_{yx}}{\partial y} + \frac{\partial \tau_{zx}}{\partial z} \right) \delta x \, \delta y \, \delta z \tag{2.20a}$$

$$\delta F_{sy} = \left(\frac{\partial \tau_{xy}}{\partial x} + \frac{\partial \sigma_{yy}}{\partial y} + \frac{\partial \tau_{zy}}{\partial z} \right) \delta x \, \delta y \, \delta z \tag{2.20b}$$

The resultant surface force can now be expressed as

$$\delta F_{sz} = \left(\frac{\partial \tau_{xz}}{\partial x} + \frac{\partial \tau_{yz}}{\partial y} + \frac{\partial \sigma_{zz}}{\partial z} \right) \delta x \delta y \delta z \tag{2.20c}$$

And

$$\delta F_s = \delta F_{sx} \hat{i} + \delta F_{sy} \hat{j} + \delta F_{sz} \hat{k} \tag{2.21}$$

And this force, combined with the body force, δF_b, yields the resultant force, δ_F, acting on the differential mass, δ_m. That is, $\delta F = \delta F_s + \delta F_b$.

2.1.2.3 Equations of Motion

The expressions for the body and surface forces can now be used in conjunction with Equation 2.17 to develop the equations of motion. In component form Equation 2.17 can be written as

$$\delta F_x = \delta m \, a_x \tag{2.22a}$$

$$\delta F_y = \delta m \, a_y \tag{2.22b}$$

$$\delta F_z = \delta m \, a_z \tag{2.22c}$$

where $\delta m = \rho \, \delta x \, \delta y \, \delta z$, and the acceleration components are given by the Eulerian method (rectangular component). It now follows that

$$\rho g_x + \frac{\partial \sigma_{xx}}{\partial x} + \frac{\partial \tau_{yx}}{\partial y} + \frac{\partial \tau_{zx}}{\partial z} = \rho \left(\frac{\partial U}{\partial t} + U \frac{\partial U}{\partial x} + V \frac{\partial U}{\partial y} + W \frac{\partial U}{\partial z} \right) \tag{2.23a}$$

$$\rho g_y + \frac{\partial \tau_{xy}}{\partial x} + \frac{\partial \sigma_{yy}}{\partial y} + \frac{\partial \tau_{zy}}{\partial z} = \rho \left(\frac{\partial V}{\partial t} + U \frac{\partial V}{\partial x} + V \frac{\partial V}{\partial y} + W \frac{\partial V}{\partial z} \right) \tag{2.23b}$$

$$\rho g_z + \frac{\partial \tau_{xz}}{\partial x} + \frac{\partial \tau_{yz}}{\partial y} + \frac{\partial \sigma_{zz}}{\partial z} = \rho \left(\frac{\partial W}{\partial t} + U \frac{\partial W}{\partial x} + V \frac{\partial W}{\partial y} + W \frac{\partial W}{\partial z} \right) \tag{2.23c}$$

Equation 2.23 shows the general differential equations of motion for a fluid. In fact, they are applicable to any continuum (solid or fluid) in motion or at rest.

2.1.2.4 Viscous Flow

To incorporate viscous effects into the differential analysis of fluid motion, one should refer to Equation 2.23. Since these equations include both stresses and velocities, there are more unknowns than equations, and therefore before proceeding it is necessary to establish a relationship between the stresses and velocities.

2.1.2.5 Stress–Deformation Relationships

For Newtonian fluids it is known that the stresses are linearly related to the rates of deformation and can be expressed in Cartesian coordinates as *for normal stresses*

$$\sigma_{xx} = -p + 2\mu \frac{\partial U}{\partial x} \tag{2.24a}$$

$$\sigma_{yy} = -p + 2\mu \frac{\partial V}{\partial y} \tag{2.24b}$$

$$\sigma_{zz} = -p + 2\mu \frac{\partial W}{\partial z} \tag{2.24c}$$

for shearing stresses

$$\tau_{xy} = \tau_{yx} = \mu \left(\frac{\partial U}{\partial y} + \frac{\partial V}{\partial x} \right) \tag{2.25a}$$

$$\tau_{yz} = \tau_{zy} = \mu \left(\frac{\partial V}{\partial z} + \frac{\partial W}{\partial y} \right) \tag{2.25b}$$

$$\tau_{zx} = \tau_{xz} = \mu \left(\frac{\partial W}{\partial x} + \frac{\partial U}{\partial z} \right) \tag{2.25c}$$

where p is the pressure, the negative of the average of the three normal stresses, which by definition is $p = -(1/3)\,(\sigma_{xx} + \sigma_{yy} + \sigma_{zz})$. For viscous fluids in motion the normal stresses are not necessarily the same in different directions.

2.1.2.6 Navier–Stokes Equations

The stresses as defined in the preceding section can be substituted into the differential equations (2.23) of motions and simplified by using the continuity equation to obtain

$$\rho \left(\frac{\partial U}{\partial t} + U \frac{\partial U}{\partial x} + V \frac{\partial U}{\partial y} + W \frac{\partial U}{\partial z} \right) = -\frac{\partial p}{\partial x} + \rho g_x + \mu \left(\frac{\partial^2 U}{\partial x^2} + \frac{\partial^2 U}{\partial y^2} + \frac{\partial^2 U}{\partial z^2} \right) \tag{2.26a}$$

$$\rho \left(\frac{\partial V}{\partial t} + U \frac{\partial V}{\partial x} + V \frac{\partial V}{\partial y} + W \frac{\partial V}{\partial z} \right) = -\frac{\partial p}{\partial y} + \rho g_y + \mu \left(\frac{\partial^2 V}{\partial x^2} + \frac{\partial^2 V}{\partial y^2} + \frac{\partial^2 V}{\partial z^2} \right) \tag{2.26b}$$

$$\rho \left(\frac{\partial W}{\partial t} + U \frac{\partial W}{\partial x} + V \frac{\partial W}{\partial y} + W \frac{\partial W}{\partial z} \right) = -\frac{\partial p}{\partial z} + \rho g_z + \mu \left(\frac{\partial^2 W}{\partial x^2} + \frac{\partial^2 W}{\partial y^2} + \frac{\partial^2 W}{\partial z^2} \right) \tag{2.26c}$$

The buoyant force ($\rho g \beta \Delta T$) can be added to Equation 2.26 as an external force if the energy equation is included.

2.1.3 First Law of Thermodynamics: The Energy Equation

2.1.3.1 Derivation of the Energy Equation

The first law of thermodynamics for a system is the time rate of increase of the total stored energy of the system equals the net time rate of energy addition by heat transfer into the system plus the net time rate of energy addition by work transfer into the system. In symbolic form, this statement is

$$\frac{D}{Dt} \int_{sys} e\rho \, dV = \left(\sum \dot{Q}_{in} - \sum \dot{Q}_{out} \right)_{sys} + \left(\sum \dot{W}_{in} - \sum \dot{W}_{out} \right)_{sys} \quad (2.27a)$$

$$\frac{D}{Dt} \int_{sys} e\rho \, dV = \left(\dot{Q}_{net\,in} + \dot{W}_{net\,in} \right)_{sys} \quad (2.27b)$$

The total stored energy per unit mass for each particle of mass in the system, e, is the summation of the internal energy per unit mass and the kinetic energy per unit mass and the potential energy per unit mass.

2.1.3.2 Differential Form

The differential form is obtained by balancing all sources of energy, conduction, convection, shear work, and the internal heat generation/dissipation. The summation of time rate of increase of the total stored energy of the system, the total conduction energy per unit volume, the total convection energy per unit volume, and the shear work per unit volume equal the heat source/sink.

$$\textit{Time rate of stored energy} = \frac{\partial}{\partial t} \rho H \quad (2.28)$$

$$\textit{Conducted energy} = \left(\frac{\partial}{\partial x} k \frac{\partial T}{\partial x} + \frac{\partial}{\partial y} k \frac{\partial T}{\partial y} + \frac{\partial}{\partial z} k \frac{\partial T}{\partial z} \right) \quad (2.29)$$

$$\textit{Convected energy} = -\left(\frac{\partial}{\partial x} \rho U H + \frac{\partial}{\partial y} \rho V H + \frac{\partial}{\partial z} \rho W H \right) \quad (2.30)$$

The shear work can be reviewed from the preceding sections. By performing the energy balance, one can obtain the differential form of the energy equation. By neglecting the work term in Equation 2.27, in air-conditioning applications, the energy equation becomes

$$\frac{\partial}{\partial x}\rho U H = \frac{\partial}{\partial x}k\frac{\partial T}{\partial x} + \frac{\partial(\Gamma \partial H/\partial x)}{\partial x} \pm S_H \tag{2.31a}$$

$$\frac{\partial}{\partial y}\rho V H = \frac{\partial}{\partial y}k\frac{\partial T}{\partial y} + \frac{\partial(\Gamma \partial H/\partial y)}{\partial y} \pm S_H \tag{2.31b}$$

$$\frac{\partial}{\partial z}\rho W H = \frac{\partial}{\partial z}k\frac{\partial T}{\partial z} + \frac{\partial(\Gamma \partial H/\partial z)}{\partial z} \pm S_H \tag{2.31c}$$

The numerical solution of heat transfer, fluid flow, and other related processes can start when the laws governing these processes have been first expressed in mathematical form amenable to solution, typically starting from the differential form of the equations. Significant theoretical advances have been made since the late 1960s and stem from the utilization of digital computers to allow the solution of simultaneous partial differential equations (PDEs), which represent the conservation of mass, momentum, energy, and species. The solution of the three-dimensional equations appropriate to a turbulent flow cannot be obtained due to limitations of storage and computing time, and as a consequence, equations based on time-averaged values of velocity, density, and temperature have been solved.

The next section has been prepared to describe and discuss the problems associated with the formulation of the equations appropriate to the airflow distribution indoors, in terms of the governing conservation equations, turbulence modeling assumptions, and computational procedure required to solve the equations with boundary and inlet conditions.

2.2 Numerical Procedure

2.2.1 Differential Equations of Motion

The partial differential equations, which govern the motion of fluids, can be expressed in tensor form as follows:

Continuity Equation
$$\frac{\partial}{\partial x_j}(\rho U_j) = -\frac{\partial \rho}{\partial t} \tag{2.32}$$

Momentum Equation
$$\frac{\partial}{\partial x_j}(\rho U_i U_j) = S_{U_i} - \frac{\partial \sigma_{ij}}{\partial x_j} - \frac{\partial \rho U_i}{\partial t} \tag{2.33}$$

where

$$\sigma_{ij} = P\delta_{ij} - \mu\left(\frac{\partial U_i}{\partial x_j} + \frac{\partial U_j}{\partial x_i}\right) + \frac{2}{3}\mu\frac{\partial U_i}{\partial x_j}\delta_{ij} \qquad (2.34)$$

δ_{ij} is the Kronecker–delta function $= \begin{cases} 0 & i \neq j \\ 1 & i = j \end{cases}$

The instantaneous velocities and densities in Equations 2.32 and 2.33 can be decomposed into mean and fluctuating components as

$$U_j = \bar{U}_j + u_j \qquad (2.35a)$$

$$\rho = \bar{\rho} + \rho' \qquad (2.35b)$$

Introducing the definitions of Equation 2.35 into the continuity equation, and then time averaging, assuming steady state, results in the equation

$$\frac{\partial}{\partial x_j}\left(\bar{\rho}\bar{U}_j + \overline{\rho'u_j}\right) = 0 \qquad (2.36)$$

Similarly, for the momentum equations,

$$\left(\bar{\rho}\bar{U}_j + \overline{\rho'u_j}\right)\frac{\partial \bar{U}_i}{\partial x_j} = \bar{S}_{U_i} - \frac{\partial \bar{\sigma}_{ij}}{\partial x_j} - \frac{\partial}{\partial x_j}\left(\bar{\rho}\,\overline{u_iu_j} + \overline{\rho'u_iu_j} + \bar{U}_j\,\overline{\rho'u_i}\right) \qquad (2.37a)$$

where

$$\bar{\sigma}_{ij} = \bar{P}\delta_{ij} - \bar{\mu}\left(\frac{\partial \bar{U}_i}{\partial x_j} + \frac{\partial \bar{U}_j}{\partial x_i}\right) + \frac{2}{3}\bar{\mu}\frac{\partial \bar{U}_i}{\partial x_j}\delta_{ij} + \frac{2}{3}\overline{\mu'\frac{\partial u_i}{\partial x_j}}\delta_{ij} - \overline{\mu'\left(\frac{\partial u_i}{\partial x_j} + \frac{\partial u_j}{\partial x_i}\right)} \qquad (2.37b)$$

Neglecting the fluctuations in laminar viscosity in Equation 2.37, the expression would take the same form, but with all variables time averaged. The equations describing the transport of scalar fluid property can, in a similar manner, be expressed as

$$\left(\bar{\rho}\bar{U}_j + \overline{\rho'u_j}\right)\frac{\partial \bar{\Phi}}{\partial x_j} = \bar{S}_\Phi - \frac{\partial \bar{J}_{\Phi,j}}{\partial x_j} - \frac{\partial}{\partial x_j}\left[\overline{\bar{\rho}u_j\varphi} + \overline{\rho'u_j\varphi} + \overline{u_j\rho'\varphi}\right] \qquad (2.38a)$$

where $\bar{J}_{\Phi,j}$ is the flux of φ along with the *j*th direction and can be expressed by Fick's law as;

$$\bar{J}_{\Phi,j} = -\Gamma_{\Phi,\ell} \frac{\partial \bar{\Phi}}{\partial x_j} \qquad (2.38b)$$

with $\quad \Gamma_{\Phi,\ell}$ equals to μ/σ_{Φ}

$\quad\quad \sigma_{\Phi}$ is the Prandtl/Schmidt number appropriate to the transport of Φ;

$\quad\quad S_{\Phi}$ is the Source/Sink of the property $\bar{\Phi}$;

and $\quad \phi$ is the fluctuating component of the property, Φ, $\phi = \Phi - \bar{\Phi}$.

The number of the time-averaged conservation equations is less than the number of the unknown terms contained in these equations.

Thus, the correlations $\overline{\rho' u_j}$, $\overline{u_i u_j}$, $\overline{u_i \varphi}$, and $\overline{\rho' \varphi}$ must be predetermined, modeled, or neglected. For nonreacting flows, the equations of interest are those governing the mass and momentum, and hence, neglecting the fluctuations of density and modeling of the velocity fluctuations is needed in order to solve the set of differential equations. Different approaches to estimate these correlation terms, which appear in Equations 2.36, 2.37, and 2.38, are considered:

- All terms involving density fluctuations may be ignored, and hence the only correlations to be modeled that contain the velocity fluctuations only. This approach is appropriate to nonreacting flows where densities are uniform or vary by a small amount.
- It is possible to write Equations 2.36, 2.37, and 2.38 in a form such that density fluctuations do not appear, and hence the equations have a form similar to that of the nonreactive equation. This involves Favre averaging (Favre, 1969) according to the equation

$$\tilde{U}_j = \bar{U}_j + \overline{\rho' u_j}/\bar{\rho} \qquad (2.39)$$

where

$\bar{U}_j$ is the average velocity

$\tilde{U}_j$ is the mass weighted velocity or Favre averaged velocity

The Favre averaged form of the conservation equations, 2.36, 2.37, and 2.38, is obtained by substituting Equation 2.39 into the conservation equations, and the result is

Continuity Equation $\qquad\qquad \dfrac{\partial}{\partial x_j} \bar{\rho} \tilde{U}_j = 0 \qquad\qquad (2.40)$

Momentum Equation $\left(\bar{\rho} \tilde{U}_j\right) \dfrac{\partial}{\partial x_j} \tilde{U}_i = \bar{S}_{U_i} - \dfrac{\partial \tilde{\sigma}_{ij}}{\partial x_j} - \dfrac{\partial}{\partial x_j} \overline{\rho u_i u_j} \qquad (2.41)$

where

$$\tilde{\sigma}_{ij} = \overline{P}\delta_{ij} - \tilde{\mu}\left(\frac{\partial \tilde{U}_i}{\partial x_j} + \frac{\partial \tilde{U}_j}{\partial x_i}\right) + \frac{2}{3}\tilde{\mu}\frac{\partial \tilde{U}_i}{\partial x_j}\delta_{ij} \qquad (2.42)$$

One of the problems associated with this method of averaging is the comparison with experimental data since most of the measurements record time-averaged properties and not the Favre averaged ones. Unfortunately, various terms in the correlation equations have themselves to be modeled, and the absolute accuracy of the results is difficult to quantify. The conservation equations are nonlinear partial differential and elliptic in form in flows where reversal flow occurs, i.e., in situations similar to those of present application. The closure of the aerodynamic equations requires the solution of supplementary equations for the shear and Reynolds stresses. The present turbulence model provides the required relationship through the eddy or turbulent viscosity concept, as described later. The velocity fluctuations that produce the shear stresses are generally expressed as

$$-\rho \overline{u_i u_j} = \Gamma_{U,t}\left(\frac{\partial \overline{U}_i}{\partial x_j} + \frac{\partial \overline{U}_j}{\partial x_i}\right) \qquad (2.43)$$

Turbulence models of this type have not been evaluated for a wide range of recirculating flows $\overline{u_i \phi}$, and prior to this work, their appropriateness to confined sudden expansion flows was unknown.

The term that appears in Equation 4.38 was replaced by the exchange coefficient $\Gamma_{\Phi,t}$ and mean gradient of Φ as

$$-\overline{u_i} = \Gamma_{\Phi,t}\frac{\partial \overline{\Phi}}{\partial x_j} \qquad (2.44)$$

The conservation equations of mass, momentum, chemical species, and energy have been discussed in general time-dependent, three-dimensional form in many references, such as Favre (1969) and Khalil (2008). In these references, attempts were made to model the effects of turbulence in terms of mean quantities, but the analysis was restricted to boundary layer and shear flow assumptions, which are not appropriate in the present flow configurations. In the present investigation, the turbulent fluctuation correlation terms, which appear in the conservation equations for confined elliptic flows, are modeled in terms of mean properties, as discussed later.

2.2.2 Turbulence Model[3]

The mean momentum equations described in Section 2.2.1 contain unknown Reynolds stress elements. In order to solve the set of momentum equations, these Reynolds stresses have to be related to the mean hydrodynamic properties of the flow field. There are many ways to relate these stresses to the mean hydrodynamic quantities, either by algebraic expression or by more complicated partial differential equations. The eddy or turbulent viscosity concept was introduced by Boussinesq,[5] who suggested the effective turbulent shear stress by a differential expression:

$$-\overline{\rho u v} \quad \text{could be replaced by} \quad \mu_t \frac{\partial \overline{U}}{\partial y} \tag{2.45}$$

The task of evaluating μ_t, the turbulent viscosity, has been the concern of many workers. A simple algebraic expression for μ_t has been proposed by Prandtl.[5]

$$\mu_t \, \alpha \left[\ell^2, \left(\partial \overline{U}_i / \partial x_j \right) \right] \tag{2.46}$$

where
  ℓ the mixing length appropriate to each flow of the boundary layer type

Alternative expressions have been noted by Launder et al.[5]
    Further advent in the evaluation of μ_t was the solution of the partial differential equation for the kinetic energy of turbulence as reported by Prandtl and Bradshaw et al.[5] The proposals of Kolmogorov,[5] Chou,[5] and Harlow et al.[5] assumed that the local state of the fluid depends on one or more turbulence quantities determined from the solution of the corresponding transport equations. In many flows this allows the turbulent flows to be characterized by two turbulence quantities, i.e., the kinetic energy of turbulence k and a characteristic length scale. The solution of differential equations for these two properties was proposed by Harlow et al.[5] and Jones and Launder.[5] A second approach to the evaluation of the Reynolds stresses is to solve transport equations for the stresses themselves, as proposed by Daly et al.[5] and Launder et al.[5] These equations represent the turbulent transport, generation, dissipation, and redistribution of Reynolds stresses.
    A two-equation turbulence model was conveniently used in the last two decades. This two-equation model is represented by equations for the kinetic energy of turbulence k and its dissipation rate ε. The turbulent viscosity μ_t is calculated from the expression

$$\mu_t = C_\mu \, \overline{\rho} \, \frac{k^2}{\varepsilon} \tag{2.47}$$

and C_μ is a constant

And the effective viscosity μ_{eff} is expressed as

$$\mu_{eff} = \mu_t + \mu \tag{2.48}$$

$$k \equiv \frac{1}{2}\overline{u_i u_i} \tag{2.49}$$

$$\varepsilon \equiv \frac{\mu}{\rho}\left(\frac{\partial u_i}{\partial x_j}\frac{\partial u_i}{\partial x_j}\right) \tag{2.50}$$

The kinetic energy k is defined as

$$\Phi = k$$

$$\Gamma_k = \frac{\mu_{eff}}{\sigma_k} \tag{2.51}$$

$$\overline{S}_k = G - \overline{\rho}\,\varepsilon \tag{2.52}$$

where
  G = Generation of turbulent energy due to mean velocity gradients

The modeled conservation equation for k was deduced, when neglecting the density fluctuations and their correlation, and can be expressed as

$$G = -\overline{\rho}\,\overline{u_i u_j}\frac{\partial \overline{U}_i}{\partial x_j} \tag{2.53}$$

And

$$\overline{\rho\,u_i u_j} = -\left[\mu_t\left(\frac{\partial \overline{U}_i}{\partial x_j} + \frac{\partial \overline{U}_j}{\partial x_i}\right) - \frac{2}{3}\left(\overline{\rho}k - \frac{\mu_t}{\overline{\rho}}\overline{U}_\ell\frac{\partial \overline{\rho}}{\partial x_\ell}\right)\delta_{ij}\right] \tag{2.54}$$

The corresponding modeled equation for the transport of ε is expressed by taking the correlations as follows:

$$\Phi = \varepsilon \tag{2.55}$$

$$\Gamma_\varepsilon = \frac{\mu_{eff}}{\sigma_\varepsilon} \tag{2.56}$$

TABLE 2.1

Model Constants

Flow	C_1	C_2	C_μ	σ_k	σ_ε
	\multicolumn{5}{c}{*k*-ε **Model Constants**}
Plane jet in a moving stream	1.44	1.92	0.09	1.0	1.3
Flow in a pipe	1.44	1.92	0.09	1.0	1.3
Wall jets on cones	1.44	1.92	0.09	1.0	1.3
Wall jets	1.44	1.92	0.09	1.0	1.3
Co-flowing jets	1.44	1.92	0.09	1.0	1.3
Flow along a twisted tape	1.44	1.92	0.09	1.0	1.3
Coaxial jets with concentration fluctuations	1.44	2.0	0.09	1.0	1.3

Source: Khalil, E. E., *Informatica*, 78, 1–7, 1977.

And

$$\overline{S}_\varepsilon = C_1 G \frac{\varepsilon}{k} - C_2 \rho \frac{\overline{\varepsilon}^2}{k} \tag{2.57}$$

C_μ, C_1, and C_2 are the constants of the turbulence model and were obtained from equilibrium flows, turbulence decay behind grids, and computer optimization, respectively. According to the recommendations of Launder et al. (1972), made after extensive examination of free turbulent flows, the constants appearing in Equations 2.47 and 2.57 take the values of Table 2.1. These constants were found appropriate for plane jets and mixing layers. Various predictions were obtained with the *k*-ε model for shear flows and were reported to be in good agreement with the corresponding measurements, e.g., Launder et al. (1972, 1973). Table 2.1 shows some of the flows that had been tested and reasonable agreement was obtained; the figures quoted are still valid and used to date in many applications.

2.2.3 Wall Functions

The form of the *k*-ε model is appropriate for high-turbulence flows with a high-turbulence Reynolds number, i.e., for values of

$$Re_W = k^{\frac{1}{2}} \ell_o / v \quad \text{greater than } 5000 \tag{2.58}$$

where

$$v \text{ is the kinematic viscosity and } \ell_o = k^{\frac{3}{2}} / \varepsilon \tag{2.59}$$

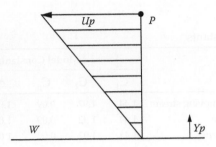

FIGURE 2.3
Near wall grid node.

In region of flow near a wall the velocities tend to zero at the wall, and hence there are zones where the local values of Re_w are so small and the viscous effect is dominant. A large number of grid nodes are needed if the momentum equations are to be solved at each node.

$$\tau_w = \left(\bar{U}_P\, y_+\right)\big/\left(y_P\, u_+\right) \tag{2.60}$$

where

$$u_+ = \frac{1}{\kappa}\ln Ey_+ \tag{2.61}$$

E and κ are constants: $E = 8.8\kappa = 0.42$

$$y_+ = y_P\,\bar{\rho}\, k_p^{1/2}\, C_\mu^{1/4}\big/\mu \tag{2.62}$$

k_p is the value of the kinetic energy of turbulence at P
and y_P is the distance normal to the wall

Near a wall, at point p shown in Figure 2.3, the flow is predominantly parallel to the wall and the shear stress is assumed constant, as reported by Launder et al.[1]

$$\frac{\bar{U}_p}{\tau_w/\rho}\, C_\mu^{1/4}\, k_p^{1/2} = \frac{1}{\kappa}\ln\left(Ey_+\right) \tag{2.63}$$

These assumptions led to an equation linking the wall shear stresses τw to the velocity parallel to the wall

$$\varepsilon_p = \left(C_\mu^{1/2}k_p\right)^{3/2}\big/ky_p \tag{2.64}$$

The constants of the logarithmic law of the wall, κ and E, depend on the wall roughness. This wall function represents the dependence of the flux of momentum to the wall on the turbulence characteristics at the point P remote from it. The value of k_p is calculated from the transport equation of κ with the diffusion of energy to the solid wall set to zero. The value of ε used to obtain k_p was assumed[1] to take the following form:

$$\frac{C_p\left(\bar{T}_p - \bar{T}_w\right)}{\left(q_w'' / \bar{\rho}\right)} C_\mu^{1/4} k_p^{1/2} = \frac{\sigma_{h,t}}{\kappa}\left[\ln Ey_+\right] + P_J \tag{2.65}$$

$$P_J = \sigma_{h,t} \frac{\pi/4}{\sin \pi/4}\left(\frac{A}{\kappa}\right)^{1/2}\left[\frac{\sigma_h}{\sigma_{h,t}} - 1\right]\left[\frac{\sigma_{h,t}}{\sigma_h}\right]^{1/4} \tag{2.66}$$

where
  $\bar{T}_w$ = wall temperature
  q_w'' = wall heat flux per unit area
  A = Van Driest's constant, 26
  $\sigma_{h,t}$ = effective Prandtl number for fully developed flow
  σ_h = laminar Prandtl number

The heat flux to the wall can be represented in a similar manner, denoting the wall temperature and heat flux per unit area.

Finally, the conservation equations are expressed in the Cartesian coordinates as follows:

$$div\left(\rho V\Phi - \Gamma_{\Phi,eff} \cdot grad\,\Phi\right) = S_\Phi \tag{2.67}$$

where
  ρ = air density, kg/m^3
  V = velocity vector
  S_Φ = source term of variable Φ
  Φ = dependent variable
  $\Gamma_{\Phi,eff}$ = effective diffusion coefficient

The effective diffusion coefficient and source term for the differential equations are listed in Table 2.2. The computational fluid dynamics (CFD) model uses approximations in calculating the turbulence quantities, such as isotropic turbulence and the Boussinesq eddy viscosity concept.

2.2.4 Numerical Solution Procedure

The conservation equations governing the flow field are difficult to solve analytically due to their complexity. With the advent of computers, it has been possible to solve them numerically. The first steps in the development

TABLE 2.2

Values of Φ, $\Gamma\Phi$, *eff*, and $S\Phi$ for Partial Differential Equations

	Φ	$\Gamma\Phi$, *eff*	S_Φ
Continuity	1	0	0
x momentum	U	μ	$-\partial P/\partial x + \rho g_x$
y momentum	V	μ	$-\partial P/\partial y + \rho g_y$
z momentum	W	μ	$-\partial P/\partial z + \rho g_z + \rho g \beta \Delta T$
H equation	H	μ/σ_H	S_H
k equation	k	μ/σ_k	$G - \rho \varepsilon$
ε equation	ε	μ/σ_ε	$C_1 \varepsilon G/k - C_2 \rho \varepsilon_2/k$

$\mu = \mu_{lam} + \mu_t$

$\mu_t = \rho \, C_\mu \, k^2/\varepsilon$

$G = \mu \, [2\{(\partial U/\partial x)^2 + (\partial V/\partial y)^2 + (\partial W/\partial z)^2\} + (\partial U/\partial y + \partial V/\partial x)^2$
$+ (\partial V/\partial z + \partial W/\partial y)^2 + (\partial U/\partial z + \partial W/\partial x)^2]$

$C_1 = 1.44$, $C_2 = 1.92$, $C_\mu = 0.09$, $\sigma_H = 0.9$, $\sigma_k = 1.0$, $\sigma_\varepsilon = 1.3$

of a numerical procedure for solving the governing differential equations are to superimpose a grid distribution on the flow domain to discretize the differential equations on all the grid points of the flow field, and to obtain equivalent algebraic expressions called finite difference equations. The accuracy of the set of the finite difference equations that approximate the partial differential equations is dependent on the formulation of the difference equations and on the number of grid nodes that represent the flow field. Three main approaches have been followed to obtain the numerical representation of the partial differential equations:

1. Taylor series expansions of the differentials with truncation of higher-order terms of the series

2. Integration over finite element employing some vibrational principles

3. Integration of the differential equations over small control volumes surrounding each grid node

As regards the method of expansion in Taylor series, the method is less general than the second and third methods because apart from offering less physical insight in the derivation of the finite difference equations, the reciprocity requirement for the fluxes at locations midway between grid nodes usually leads to the central difference formulation, which is an inaccurate representation of the fluxes at high Peclet numbers. Therefore, this method was not used. The second method is associated with the finite element technique and was discarded in the present work because of the difficulty of describing the boundary conditions. Thus, the third method was used to obtain the finite difference equations for the conservation of mass, momentum, and total enthalpy.

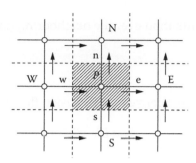

FIGURE 2.4
Finite difference grid.

2.2.5 Finite Difference Equations

The finite difference counterpart of the general partial differential equations is derived by supposing that each variable is enclosed in its own control volume, as shown in Figure 2.4. The grid node P was surrounded by two X direction neighboring points W and E in the west and east directions and two nodes N and S in the north and south Y directions. The partial differential equation is integrated over the control volume with the aid of Φ and the rates of generation/destruction of the entity Φ within the cell and its transport by convection and diffusion across the cell boundaries. The former is represented in linearized form as

$$S_\Phi = \int_V S_\Phi \, dv = S_u + S_P \, \Phi_P \tag{2.68}$$

and the latter by expressions of the form

$$\rho U_w \left[(\Phi_P + \Phi_W)/2\right] A - \Gamma_{\Phi_w} \left[(\Phi_P - \Phi_W)/\delta x_{PW}\right] A \tag{2.69}$$

where the quantity $Pe_W \equiv \rho U_w \, \delta x_{PW}/\Gamma_{\Phi_w}$ is small and by:

$$\rho U_w \, \Phi_W, U_w > 0; \quad \rho U_w \, \Phi_P, U_w < 0 \tag{2.70}$$

when Pe_w is large

Here the subscripts P and W refer to the central and west nodes, respectively, and w denotes the intervening cell boundary. Assembly of the above and similar expressions for the remaining boundaries yields the finite difference equation as

$$(A_P - S_P)\Phi_P = \sum_n A_n \Phi_n + S_u \tag{2.71}$$

where $\sum_n$ denotes summation over the neighboring nodes, N, S, R, L, E, and W, $A_P = \sum_n A_n$

The coefficient A_n is the net convection diffusion flux and is expressed as

$$A_n = D_n^* - C_n \qquad (2.72)$$

where

$$D_n^* = 0.5\left(D_n + |C_n| + |D_n - |C_n||\right) \qquad (2.73)$$

$$D_n = \Gamma_{\Phi,n} A/\delta x \qquad (2.74)$$

$$C_n = \rho U_n A/2 \qquad (2.75)$$

Equations of this kind are written for each of the variables U, V, W, and Φ at every cell, with appropriate modifications being made to the total flux expressions (2.69 and 2.70) at cells adjoining the boundaries of the solution domain to account for the conditions imposed there. An equation for the remaining unknown, pressure, is obtained by combining the continuity and momentum in the manner explained in Patankar et al.[4] This entails connecting changes in pressure, denoted P', with changes in the velocities U, V, and W by approximate formulae derived from the momentum finite difference equations as

$$U_e = U_e^* + D_E^u\left(P_P' - P_E'\right) \qquad (2.76)$$

and

$$U_w = U_w^* + D_W^u\left(P_W' - P_P'\right) \qquad (2.77)$$

and similarly for all velocities components.

The starred values are guessed values and Du is the pressure difference coefficient. A substitution of these formulae into the continuity equation yields a partial differential equation for p' similar to Equation 2.71, with Su now representing the local continuity imbalance in the prevailing velocity field.

2.2.6 Solution Algorithm

The finite difference equations are solved by iteration, employing inner and outer sequences. The outer iteration sequence involves the cyclic application of the following steps: First, a field of intermediate velocities, denoted by starred values, is obtained by solving the associated momentum equations using the prevailing pressures p^*. Then, continuity is enforced by solving the equations for p' and thereby determining the required adjustments to the velocities and pressure. The equations for the remaining variables are solved in turn, and the whole process is repeated until a satisfactory solution is obtained as the residuals in any of the finite difference equations are less than 0.1% for flow, 1% for k and ε, and 0.1% for energy. The inner iteration sequence is employed to solve the equation sets for the individual variables. Solution is by a form of block iteration, in which a simple recurrence formula is used to solve simultaneously for the Φs along each grid line, in the line-by-line counterpart of point Gauss Seidel iteration. Complete convergence of the inner sequence is not necessary, and usually one to three applications of block procedure suffices.

The numerical solution is required to pass two acceptance tests. First, it must satisfy the finite difference equations when substituted into them; typically, the imbalance must be 0.1% or less. Second, it must be invariant with further increases in the number of grid nodes. These two tests are discussed as follows.

2.2.6.1 Convergence and Stability

The simultaneous and nonlinear character of the finite difference equations necessitates that special measures are employed to procure numerical stability (convergence): these include underrelaxation of the solution of the momentum and turbulence equations by underrelaxation factors that relate the old and new values of Φ as follows:

$$\Phi = \gamma\, \Phi_{new} + (1 - \gamma)\, \Phi_{old} \qquad (2.78)$$

Here γ denotes the underrelaxation factor, which was varied between 0.3 and 0.5 for the three velocity components as the number of iterations was increased. For the turbulence quantities, γ was taken as 0.2, and for other variables, as 0.3. The detailed information about the required iterations for convergence and iteration time should be highlighted in each case as proof of attaining the final correct predictions.

2.3 Conclusions

This chapter discusses the physical phenomena prevailing in most engineering problems, and presents the engineering requirements and their thoughts to achieve the most appreciated tools for solving their problems. It introduces the philosophy of mathematical modeling with regard to other approaches to solving the engineering problems, and introduces the modeling ability to solving the physical problems, with its ability in forecasting and the limits to its success. The present chapter also introduces some of the new areas that can be investigated using the modeling methods and the appropriate techniques to handle each phenomenon. It introduces technical recommendations for mathematical model selection for different types of fluid and thermal applications. It also introduces different experienced recommendations to optimize the stability of the computation procedure and enhance the software development. This is complemented with definitions of suitable and optimum strategies to achieve optimum CFD programs and suitable engineering software.

References

1. Khalil, E.E. 1978. Numerical procedures as a tool to engineering design. *Informatica* 78, 1–7.
2. Schlichting, H. 1968. *Boundary Layer Theory*. New York: McGraw-Hill.
3. Khalil, E.E. 1982. *Modeling of Furnaces and Combustors*. United Kingdom: Abacus Press.
4. Patankar, S.V. and Spalding, D.B. 1974. A calculation procedure for heat, mass, and momentum transfer in three dimensional parabolic flows. *Int. J. Heat & Mass Transfer* 15, 1787–1799.
5. Launder, B.E., and Spalding, D.B. 1974. The numerical computation of turbulent flows. *Computer Methods App. Mech.* 3(2), 269–289.

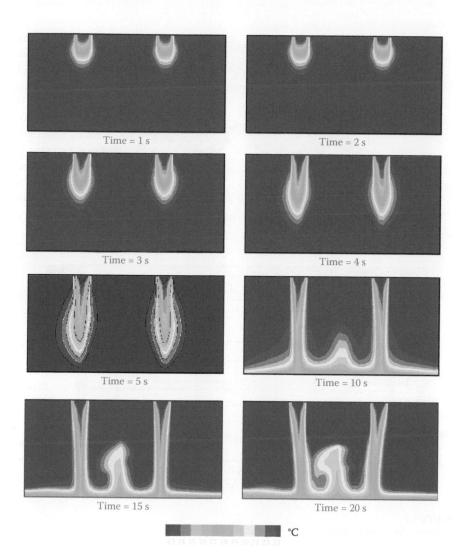

Time = 1 s

Time = 2 s

Time = 3 s

Time = 4 s

Time = 5 s

Time = 10 s

Time = 15 s

Time = 20 s

°C

13 14 15 16 17 18 19 20 21 22 23

EXAMPLE 1.3

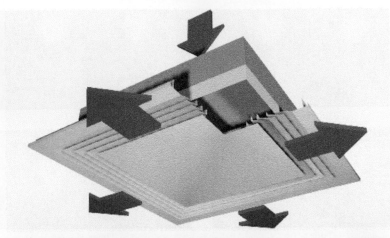

Hr	Lr/Wr	Hs	ΔT	qv	a	....			
[m]	[m]	[m]	[°C]	[l/s]		[m/s]	[m/s]	[m/s]	[m/s]
3.0	8.0	3.0	−8	110.0	0.0	0.27	0.20	0.40	0.60

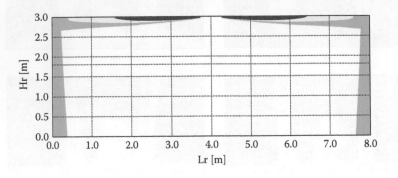

FIGURE 1.7
Ceiling architectural diffuser.

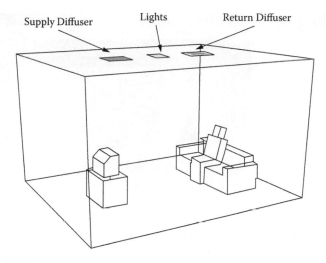

FIGURE 3.1
Case geometry.

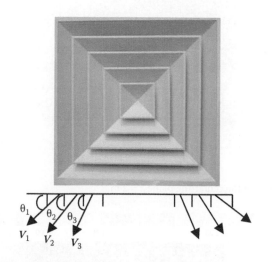

FIGURE 3.2
Supply diffuser.

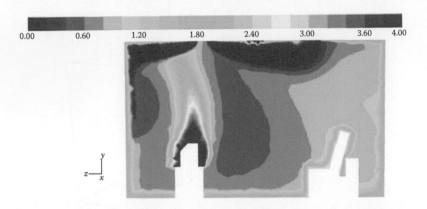

FIGURE 3.12
PMV contours at the centerline plane $x = 2$ m.

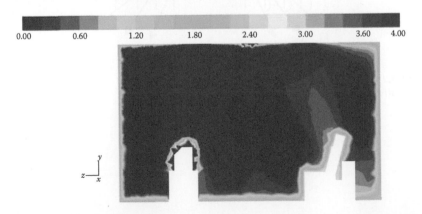

FIGURE 3.15
Temperature contours at the centerline plane $x = 2$ m.

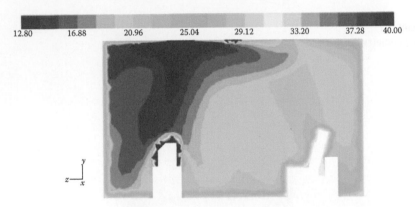

FIGURE 3.19
PMV contours at the centerline plane $x = 2$ m.

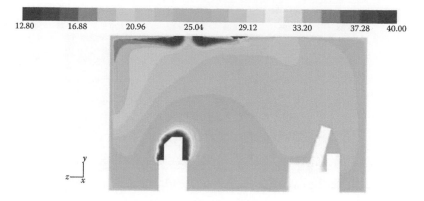

FIGURE 3.22
Temperature contours at the centerline plane $x = 2$ m.

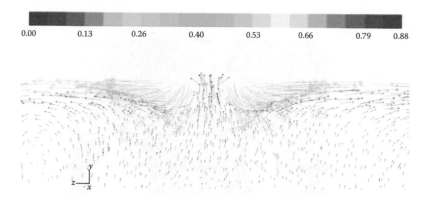

FIGURE 3.23
Velocity vectors of air supplied from the circular diffuser.

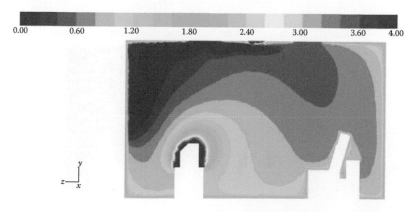

FIGURE 3.25
PMV contours at the centerline plane $x = 2$ m.

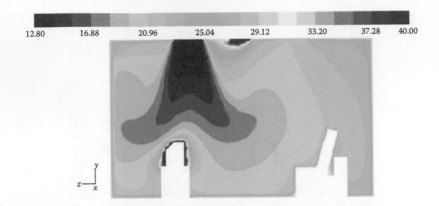

FIGURE 3.29
Temperature contours at the centerline plane $x = 2$ m.

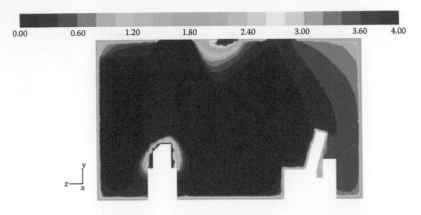

FIGURE 3.31
PMV contours at the centerline plane $x = 2$ m.

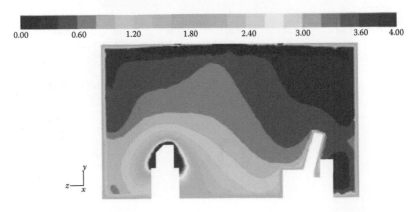

FIGURE 3.37
PMV contours at the centerline plane $x = 2$ m.

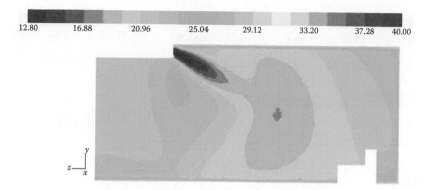

FIGURE 3.41
Temperature contours in a vertical plane at $x = 1.2$ m.

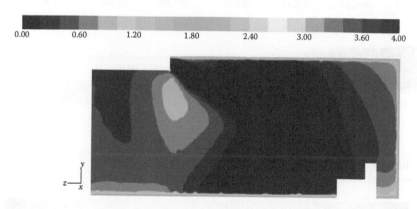

FIGURE 3.43
PMV contours in a vertical plane at $x = 1.2$ m.

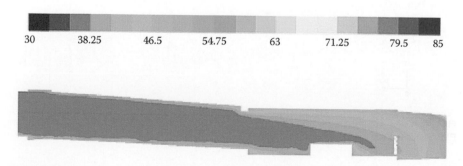

FIGURE 4.6
RH% contours for a mid-plane along the KV1 tomb axis.

| 30 | 38.25 | 46.5 | 54.75 | 63 | 71.25 | 79.5 | 85 |

FIGURE 4.8
RH% contours for a mid-plane along the KV9 tomb axis.

Effect the outside air condition has on the temperature patterns inside the tombs

| 298.00 | 300.40 | 302.81 | 305.21 | 307.62 | 310.00 | 312.40 | 314.00 |

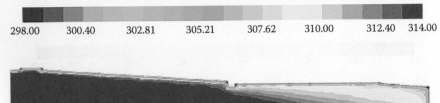

FIGURE 4.10
Temperature contours, K, for mid-plane, $z = 1.8$ m, KV1.

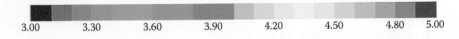

| 3.00 | 3.30 | 3.60 | 3.90 | 4.20 | 4.50 | 4.80 | 5.00 |

FIGURE 4.12
PMV contours for a mid-plane at $z = 1.8$ m, KV1.

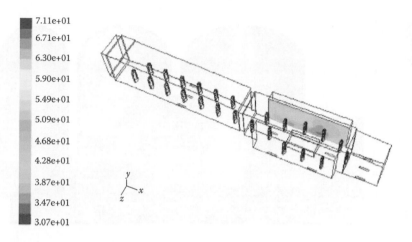

7.11e+01	
6.71e+01	
6.30e+01	
5.90e+01	
5.49e+01	
5.09e+01	
4.68e+01	
4.28e+01	
3.87e+01	
3.47e+01	
3.07e+01	

FIGURE 4.29

Relative humidity % near wall at KV1 with proposed design (24 visitors). (From Khalil, E. E., *Proceedings of ASHRAE RAL*, Athens, Greece, 2005.)

FIGURE 4.31

Simulated relative humidity contours on wall artifacts in KV1. (From Khalil, E. E., and Abdelaziz, O. O. A., *Fluent News*, 28, 2006.)

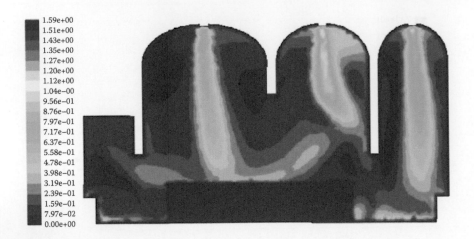

■	1.59e+00
	1.51e+00
	1.43e+00
	1.35e+00
	1.27e+00
	1.20e+00
	1.12e+00
	1.04e−00
	9.56e−01
	8.76e−01
	7.97e−01
	7.17e−01
	6.37e−01
	5.58e−01
	4.78e−01
	3.98e−01
	3.19e−01
	2.39e−01
	1.59e−01
	7.97e−02
■	0.00e+00

FIGURE 5.9
Contours of velocity at $x = 4$.

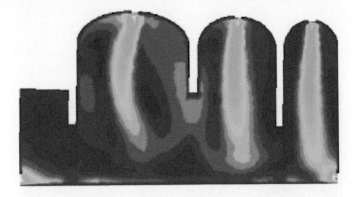

FIGURE 5.10
Contours of velocity at $x = 15$ m.

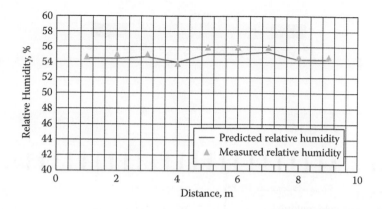

FIGURE 5.25
Measured and predicted humidity percentage at 1.0 m above floor in the church. (From Khalil, E. E., Flow Regimes and Heat Transfer Patterns in Archeological Climatized Church of Christ, Cairo, IMECE-2012-85088, *Proceedings of ASME 2012 International Mechanical Engineering Congress and Exposition*, Houston, TX, 2012.)

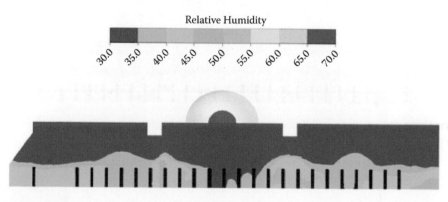

FIGURE 5.32
Predicted relative humidity contours (%).

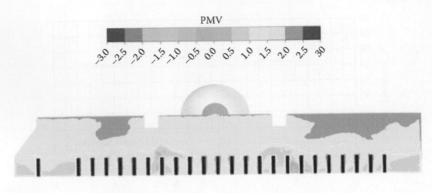

FIGURE 5.33
Predicted mean vote contours.

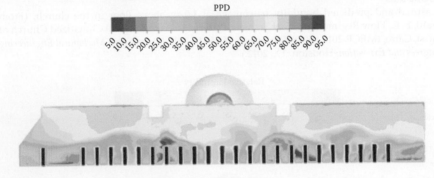

FIGURE 5.34
Predicted percentage dissatisfied contours (%).

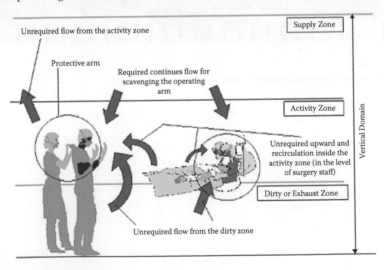

FIGURE 6.1
Airflow nature in the surgical operating theaters.

3

Airflow Regimes and
Thermal Comfort in a Room

A computational fluid dynamics model is developed to examine the airflow characteristics of a room with different supply air diffusers. This chapter is devoted to numerically investigating the influence of location and number of air supply and extract openings on airflow properties in a typical seating room. The work focuses on airflow patterns and the thermal behavior in the room with a few occupants. As an input to the full-scale 3D room model, a 2D air supply diffuser model that supplies direction and magnitude of airflow into the room is developed. The air distribution effect on thermal comfort parameters was investigated depending on changing the air supply diffuser type, angles, and velocity. Air supply diffuser locations and number were also investigated.

The preprocessor Gambit is used to create the geometric model with parametric features. Commercially available simulation software Fluent 6.3[6] is incorporated to solve the differential equations governing the conservation of mass, three momentum, and energy in the processing of airflow distribution.[1-10] Turbulence effects of the flow are represented by the well-developed two-equation turbulence model. In this chapter, the so-called standard k-ε turbulence model, one of the most widespread turbulence models for industrial applications, was utilized. Basic parameters included in this work are air dry-bulb temperature, air velocity, relative humidity, and turbulence parameters that are used for numerical predictions of indoor air distribution and thermal comfort.

The thermal comfort predictions through this work were based on the predicted mean vote (PMV) model and the percentage people dissatisfied (PPD) model; the PMV and PPD were estimated using Fanger's model. Throughout the investigations, numerical validation is carried out by way of comparisons of published experimental results whenever available. Good qualitative agreement was generally observed.

3.1 Numerical Model

3.1.1 Case Geometry

The present chapter is concerned with the airflow patterns in a simple room due to changing the supply diffuser shape and angles. The room's main

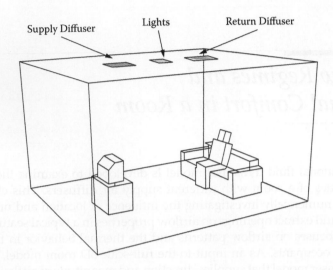

FIGURE 3.1
(See color insert.) Case geometry.

dimensions are 4 m width, 5 m length, and 3 m height, as shown in Figure 3.1. A person seated on a sofa is modeled, and a television is added in front of him. The air is supplied to the room with different conditions, as shown later in the next section.

3.1.2 Boundary Conditions

This section will present the various boundary conditions assumed for all the studied cases herein described.

3.1.2.1 Walls

The room walls were set in the solver to be a constant temperature surface, and its temperature was assumed to be 30°C. The wall material was simulated to be gypsum, which matches the real configuration wall properties.

3.1.2.2 Interior

The room furniture and non-heat-dissipating equipment in the room were set to be at zero wall heat fluxes.

3.1.2.3 Supply Grilles

The turbulence intensity was assumed to be 5% based on manufacturer's catalog data; see Figure 3.2 for an example and details given in Table 3.1.

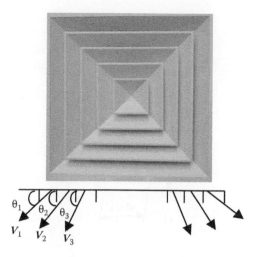

FIGURE 3.2
(See color insert.) Supply diffuser.

TABLE 3.1

Inlet Air Conditions

Case	T_{supply}	θ_1	θ_2	θ_3	V_{supply}
1	12.8	15	30	60	0.9
2	12.8	15	30	60	0.65
3	12.8	30	60	90	0.65

3.1.2.4 Light Heat Load

Lighting fixtures are mounted at the ceiling; the light heat flux is set to a value of 555 w/m² for an area of 0.18 m².

3.1.2.5 Human Body

3.1.2.5.1 A Seated Man's Dimensions

Figure 3.3 shows the configuration and dimensions of the assumed model of the seated occupant.

3.1.2.5.2 Body as a Heat Source

The body is treated as a wall at a constant temperature, and it is set according to Figure 3.4, where the skin temperature is a function of the metabolic rate in Met (1 Met = 58 W/m²). As it has been assumed that the occupant's metabolic rate is 116 W/m² (2 Met), this is equivalent to 32.5°C skin temperature, and the body is assumed to have zero diffusive flux.

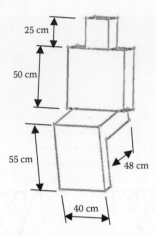

FIGURE 3.3
Configuration of occupant's model.

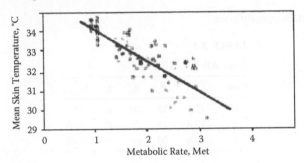

FIGURE 3.4
Mean skin temperature as a function in activity level. (From Fluent, Inc., Fluent 6.2, documentation, 2010.)

3.1.2.6 Return Grilles

Set to be a pressure outlet with the same water mass fraction, the return grilles have the same dimensions as the supply ones, and the turbulence intensity is assumed to be 3%. Pressure outlet boundary conditions were prescribed on all the return grilles of the surrounding air. This is a typical boundary condition type for flows that can reverse direction near a boundary. It requires the specification of a static pressure and temperature of backflow at the outlet boundary. Since no extraction fans are mounted in the return air duct, these parameters were set to 101.325 Pa and 292 K, respectively.

3.1.2.7 Television Heat Load

Television is placed on a table in the middle of the room, and the heat load is assumed to be 200 W, which corresponds to a heat flux of 200 w/m² for an area of 1 m².

3.2 Flow Pattern and Thermal Behavior

3.2.1 General

This part shows the parametric studies done on the room model presented in Chapter 2. The first case shows the best results in room comfort, as shown in its PPD histogram, without paying attention to energy efficiency. In the second case the airflow is changed in order to enlarge PPD differences in other cases. Other designs show different air distributions depending on changing diffuser angles, number, type, and location.

3.2.2 Design Alternatives

3.2.2.1 Design 1

Boundary conditions for this case and diffuser angles were mentioned in Chapter 2. Temperature contours at selected planes are shown, as well as the PMV, the velocity vectors at the supply diffuser, and the PPD histogram. As shown in Figure 3.5 for a vertical plane in the middle of the room, the temperature is 40°C near the heat sources (lamp and television). The temperature around the sofa is 24°C. Figure 3.6 presents a prediction of the thermal pattern at a plane 1.0 m above the room floor, identifying the various thermal zones around the seating area to the right of the room and around the TV set.

Figure 3.7 presents the predicted PMV for thermal comfort showing values on the order of +1.2 near the human. This is better highlighted in Figure 3.8, which shows the room at 1 m above the finished floor to be within comfort levels.

The PPD is shown in Figure 3.9 to consolidate the values of Figure 3.8. In these figures different ceiling outlets are to be numerically tested to evaluate the level of comfort that these can provide to the occupants of the room.

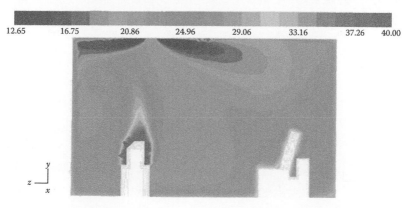

FIGURE 3.5
Temperature contours at the centerline plane $x = 2$ m.

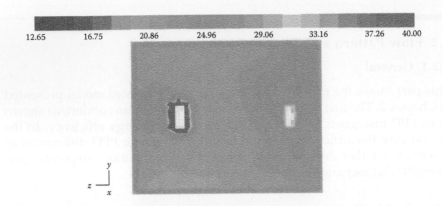

FIGURE 3.6
Temperature contours of a horizontal plane at $y = 1$ m.

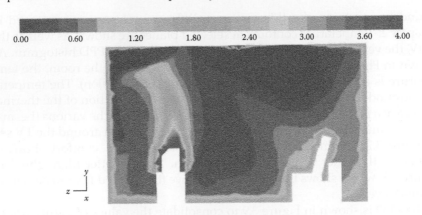

FIGURE 3.7
PMV contours at the centerline plane $x = 2$ m.

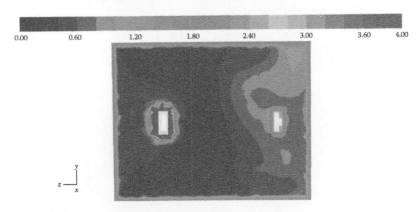

FIGURE 3.8
PMV contours in a horizontal plane at $y = 1$ m.

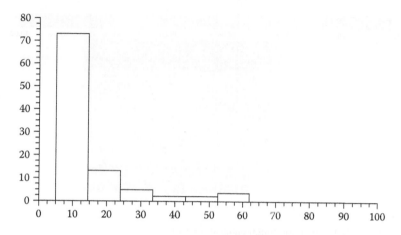

FIGURE 3.9
Room PPD histogram.

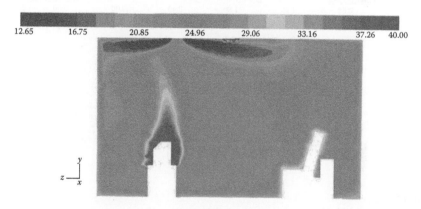

FIGURE 3.10
Temperature contours at the centerline plane $x = 2$ m.

3.2.2.2 Design 2: Square Diffuser (60–30–15°)

The basic geometrical characteristics of this square ceiling diffuser were fed into the boundary and inlet conditions; Figures 3.10 to 3.13 demonstrate the predicted temperature contours PMV and PPD at vertical and horizontal planes. These generally demonstrated higher temperatures in the vicinity of the occupant relative to Figure 3.5. This is seen also by the less intense emerging cold air jet from the ceiling. Figure 3.14 shows higher values of dissatisfaction.

3.2.2.3 Design 3: Square Diffuser (90–60–30°)

This design utilizes different diffuser angles and results in the thermal patterns and comfort levels indicated in Figures 3.15 to 3.18.

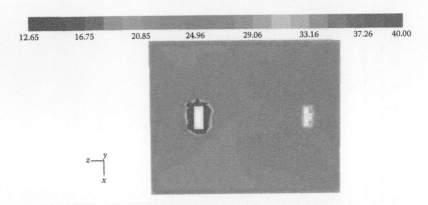

FIGURE 3.11
Temperature contours in a horizontal plane at $y = 1$ m.

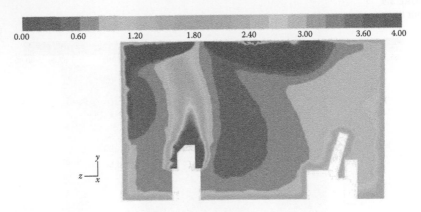

FIGURE 3.12
(**See color insert.**) PMV contours at the centerline plane $x = 2$ m.

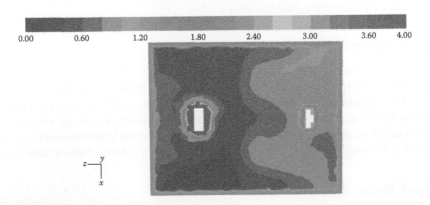

FIGURE 3.13
PMV contours in a horizontal plane at $y = 1$ m.

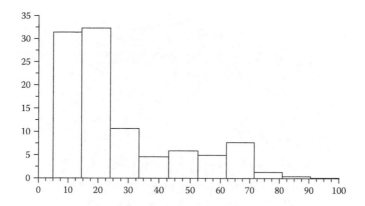

FIGURE 3.14
Room PPD histogram.

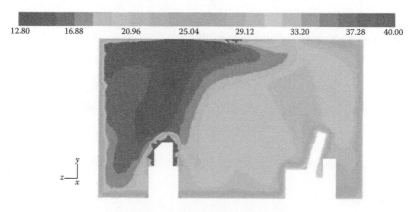

FIGURE 3.15
(**See color insert.**) Temperature contours at the centerline plane $x = 2$ m.

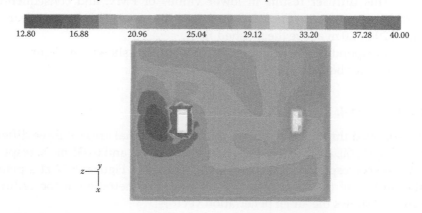

FIGURE 3.16
Temperature contours in a horizontal plane at $y = 1$ m.

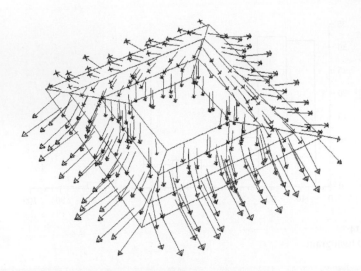

FIGURE 3.17
Velocity vectors of air supplied from a square diffuser.

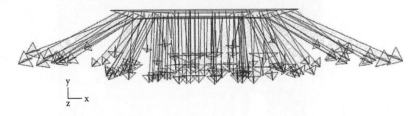

FIGURE 3.18
Supplied air angles.

The corresponding thermal comfort parameters are shown in Figures 3.19 to 3.21. This diffuser results in lower values of PMV, and consequently toward a greater feeling of coolness all over the room. This is very clear in Figure 3.20 at 1.0 m above the finished floor.

The corresponding percentages dissatisfied are shown in Figure 3.21, which indicates better feeling of comfort in the room.

3.2.2.4 Design 4: Circular Diffusers (90–60–30°)

Air is supplied through three annular areas with equal areas at three different angles, 90, 60, and 30°, with discharges 0.025, 0.6, and 0.035 m³/s, respectively. The corresponding predictions are shown in Figures 3.22 at a plane at the middle of the room showing smaller jet penetration at the ceiling. Figure 3.23 shows the air jet penetration vectors.

Figure 3.24 presents the thermal pattern at 1.0 m above the finished floor and temperatures are 22°C around the seating area.

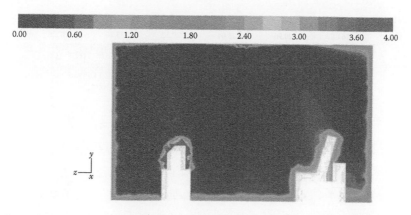

FIGURE 3.19
(See color insert.) PMV contours at the centerline plane $x = 2$ m.

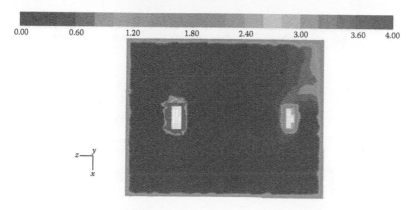

FIGURE 3.20
PMV contours of a horizontal plane at $y = 1$ m.

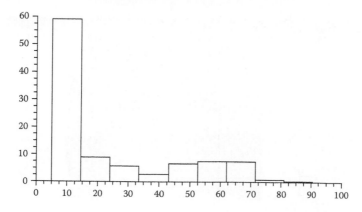

FIGURE 3.21
Room PPD histogram.

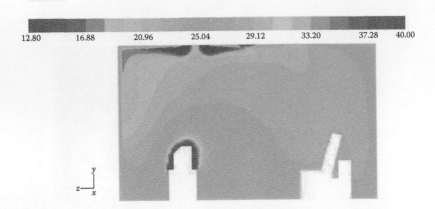

FIGURE 3.22
(See color insert.) Temperature contours at the centerline plane $x = 2$ m.

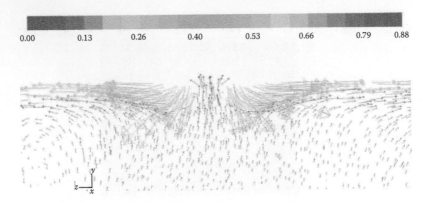

FIGURE 3.23
(See color insert.) Velocity vectors of air supplied from the circular diffuser.

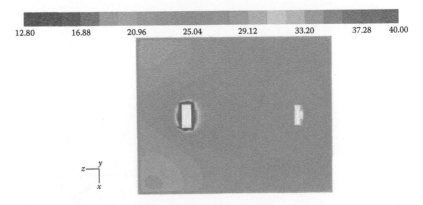

FIGURE 3.24
Temperature contours in a horizontal plane at $y = 1$ m.

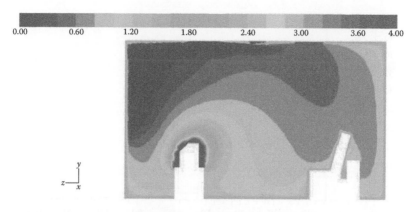

FIGURE 3.25
(See color insert.) PMV contours at the centerline plane $x = 2$ m.

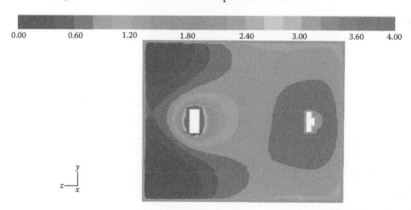

FIGURE 3.26
PMV contours of a horizontal plane at $y = 1$ m.

Predictions of PMV shown in Figures 3.25 and 3.26 indicated different behaviors than those shown earlier, and lower PMVs are generally prevailing, and these are around +0.8, which is fine for comfort.

The corresponding PPDs are shown in Figure 3.27, and once more, less dissatisfaction is generally acknowledged with this design. So it can be easily recommended to use such a design for rooms.

3.2.2.5 Design 5: Swirl Flow

A different ceiling diffuser of the swirl type design is suggested here. In this case, the air is supplied at a flow rate of 0.12 m³/s, but at two equal components in the axial and tangential directions, as outlined in Figure 3.28, which shows the vectorial velocity components emerging from the diffuser into the room.

The corresponding predictions of the thermal contours at a vertical plane in the middle of the room at a plane of $X = 2$ m are shown in Figure 3.29. The

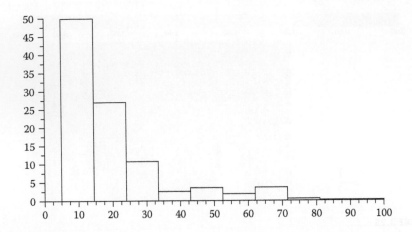

FIGURE 3.27
Room PPD histogram.

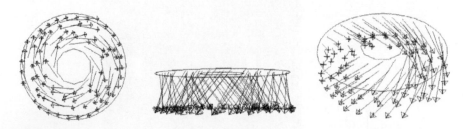

FIGURE 3.28
Grille airflow directions.

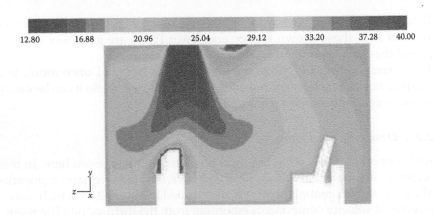

FIGURE 3.29
(See color insert.) Temperature contours at the centerline plane $x = 2$ m.

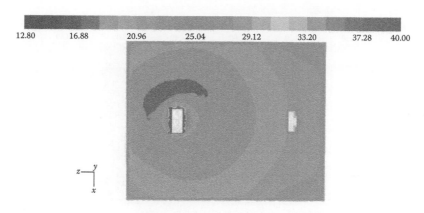

FIGURE 3.30
Temperature contours in a horizontal plane at $y = 1$ m.

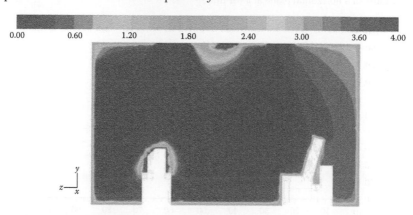

FIGURE 3.31
(**See color insert.**) PMV contours at the centerline plane $x = 2$ m.

jet penetration is well identified, and the drop may reach 1 m above the floor. The circular effect of the jet is persistent and shows at even 1 m above the room floor, as also shown in Figure 3.30.

Looking at Figure 3.31, one can visualize the lower values of PMV in the room that can be expected due to the rotational motion of the swirl diffuser that aids better distribution of the cold air into the room, resulting in better mixing and even some feeling of coldness. Figure 3.32 confirms that finding, as it outlines the PMV at 1.0 m above the floor to be within +0.4.

The corresponding percentage dissatisfaction shown in Figure 3.33 consistently indicates that this is a better design.

3.2.2.6 Design 6: Side Return

In some design configurations, the return air is typically side wall mounted, and the supply is in the middle of the room, as shown in Figure 3.34. In this

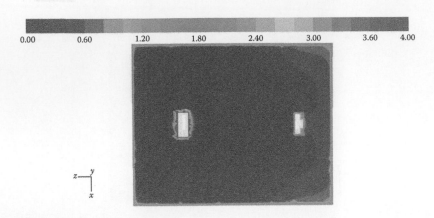

FIGURE 3.32
PMV contours of a horizontal plane at $y = 1$ m.

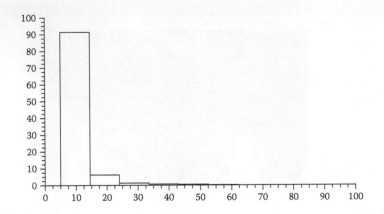

FIGURE 3.33
Room PPD histogram.

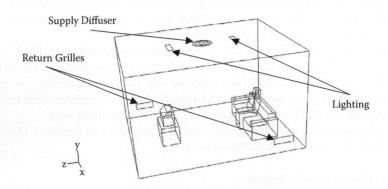

FIGURE 3.34
Room configuration.

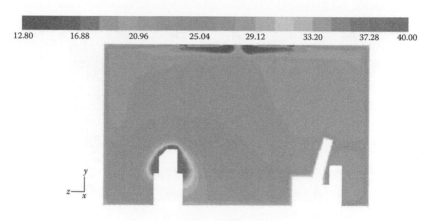

FIGURE 3.35
Temperature contours at the centerline plane $x = 2$ m.

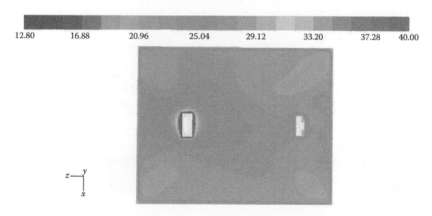

FIGURE 3.36
Temperature contours in a horizontal plane at $y = 1$ m.

case the air is supplied from the central circular diffuser, and is extracted through the side rectangular grilles. Air is supplied at 0.12 m³/s.

The thermal pattern predicted for that design indicated higher temperature in the occupied area of the room, at least 2° higher than those predicted for the earlier designs (see Figure 3.35). The temperature contours at 1 m above the floor, shown in Figure 3.36, also indicated a higher average temperature, indicating poorer mixing in the room. Nevertheless, the predicted PMV values shown in Figures 3.37 and 3.38 are of the order of +1.0 to 1.2. In many design situations the Mechanical and Electrical Projects Consultant (MEP) is confronted with the architectural constraints to have return air grilles in the ceiling for some different reasons; these figures indicate that one expects somewhat poorer comfort levels.

This PPD, shown in Figure 3.39, indicated rather good comfort, but worse than that of Figure 3.33 for the swirl diffuser.

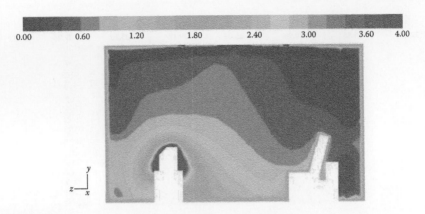

FIGURE 3.37
(See color insert.) PMV contours at the centerline plane $x = 2$ m.

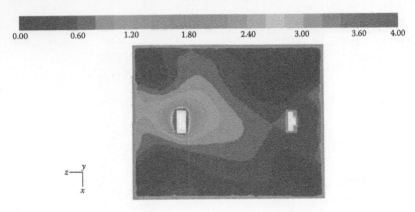

FIGURE 3.38
PMV contours of a horizontal plane at $y = 1$ m.

3.2.2.7 Design 7: Side Supply

In this case the air is supplied from a side grille, as shown in Figure 3.40. This case represents some of the actual hotel room designs. The air is kept at a discharge of 0.12 m³/s. Air is making a 30° angle with the negative Z direction as it enters the room.

Figure 3.41 presents the air temperature contours as predicted with the numerical algorithm solving the governing equations of mass, momentum, and energy in a full three-dimensional configuration with more than 2 million grid nodes. The cold air jet penetration is clear and is taken at an angle of 30°, as shown. The jet through and drop are naturally controlled by the diffuser design and geometry. The real temperature effect at the occupant is 22°C, acceptable for temperature comfort factor. The temperature distributions at 1 m above the floor are not homogeneous, and temperature differ-

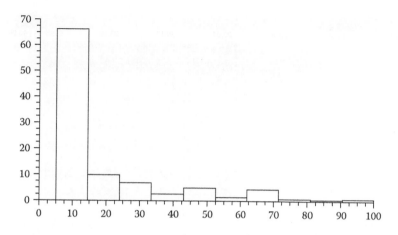

FIGURE 3.39
Room PPD histogram.

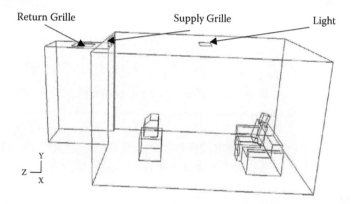

FIGURE 3.40
Room configuration.

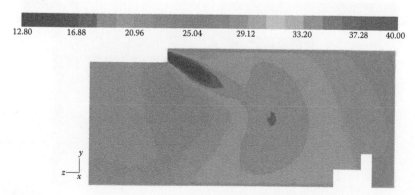

FIGURE 3.41
(See color insert.) Temperature contours in a vertical plane at $x = 1.2$ m.

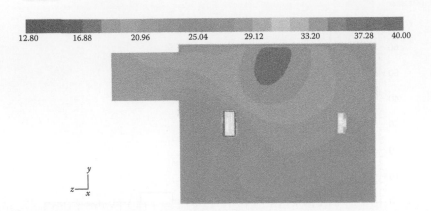

FIGURE 3.42
Temperature contours in a horizontal plane at $y = 1$ m.

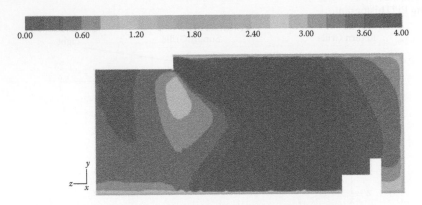

FIGURE 3.43
(**See color insert.**) PMV contours in a vertical plane at $x = 1.2$ m.

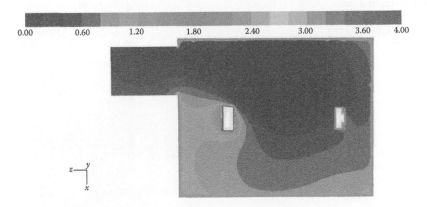

FIGURE 3.44
PMV contours of a horizontal plane at $y = 1$ m.

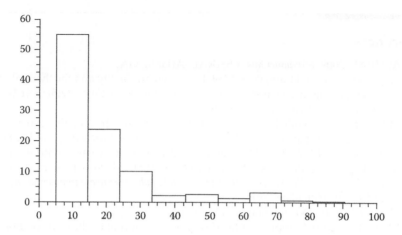

FIGURE 3.45
Room PPD histogram.

ences across the room can handsomely vary by more than 8°, as indicated in Figures 3.41 and 3.42.

The corresponding predictions of the contours of the PMV and PPD histograms are shown in Figures 3.43 to 3.45, respectively. The room conditions indicated that the values of PMV are within +0.4 and 1.2 in the room at 1.0 m above the finished floor level.

3.3 Conclusions

Numerical computational procedures can successfully yield very important and detailed information regarding the airflow and thermal patterns in rooms as simple as those shown here. Details of room furniture, occupants, and equipment can be successfully represented. Various diffusers feeding cold air to the room were analyzed in terms of thermal patterns and comfort index. Different square diffusers were incorporated with supply angles of 15-30-60° and 30-60-90°, and the 30-60-90° angles showed better results, as shown before in the PPD histograms. Circular diffusers didn't show significant improvement compared to square diffusers.

The proposed circular ceiling swirl diffusers showed the best results, as 90% of the room volume has a PPD of 10%.

The design cases with side supply air grilles were investigated, as they represent a common design feature in many hotel and individual office rooms applications, but the obtained results did not match those of the superior ceiling diffuser designs.

References

1. ASHRAE. 2009. *Fundamentals*. ASHRAE, Atlanta, GA.
2. Berglund, L. G., and Cain, W. S. 1989. Perceived air quality and the thermal environment, the human equations: Health and comfort. *Proceedings of ASHRAE/SOEH Conference IAQ'89*, Atlanta, GA, pp. 93–99
3. ElBialy, E., and Khalil, E. E. 2012. Air flow distribution effect on thermal comfort in a living room. Paper AIAA-2012-4187. Presented at Proceedings of IECEC.
4. ElBialy, E., and Khalil, E. E. 2013. Air flow regimes and thermal comfort in a living room. Paper 9881. Presented at Proceedings of ASHRAE 2013, January 2013.
5. Fanger, P. O. 1972. *Thermal comfort: Analysis and application in environmental engineering*. McGraw-Hill, New York.
6. Fluent 6.2. 2010. Documentation. Fluent, Inc.
7. Khalil, E. E. 2000. Computer aided design for comfort in healthy air conditioned spaces. *Proceedings of Healthy Buildings 2000*, Finland, 2, 461–467.
8. Khalil, E. E. 2006. Preserving the tombs of the pharaohs. *ASHRAE Journal* 48(12), 34–38.
9. Khalil, E. E. 2012. CFD history and applications. *CFD Letters* 4(2), 1–4.
10. Olesen, B. W. 2000. Guidelines for comfort. *ASHRAE Journal* 5, 41–46.

4

Airflow Regimes and Thermal Pattern in Archeological Monuments

The tombs of the kings in Valley of the Kings, Luxor, are considered to be one of the tourism industry's bases in Egypt due to their uniqueness all over the world. Hence, they should be preserved from the different factors that might cause harm to their wall paintings. One of these factors is the excessive relative humidity (RH), as it increases the bacteria and fungus activity inside the tomb, in addition to its effects on the mechanical and physical properties of materials. This chapter aims to provide controlled climate to the tombs of the Valley of the Kings, with complete monitoring of air properties, temperature, relative humidity, and carbon oxides, and air quality parameter mechanical distributions inside selected tombs of the Valley of the Kings that are listed in Appendix A. A complete climate control and monitoring of air will be affected with the aid of a mechanical ventilation system extracting air at designated locations in the wooden raised floor of the tombs. The location, size, and extracted air are to be predicted and optimized by use of computational fluid dynamics (CFD) software. The CFD modeling techniques would solve the continuity, momentum, energy, and species transport equations in addition to k-ε model equations for turbulence closure. The SIMPLEC algorithm is used for the pressure-velocity coupling, and a second-order upwind scheme is used for discretization of the governing equations. Mesh sizes used in the present work exceeded 1,000,000 mesh volumes to adequately represent the flow characteristics at various locations.

Throughout this chapter, the outside air conditions, number of visitors, and airside system design effect on the tombs' airflow characteristics are investigated in order to reach the optimum ventilation design as well as the favorable working conditions for a particular tomb. The mechanical engineering design of the airflow system, including under raised floor flexible duct routings, sizes, grille locations, etc., will be presented with professional design drawings to be used as bases for execution and installation. A permissible number of simultaneous visitors for each tomb should be made in order to limit the relative humidity inside any given tomb.

Basic requirements include the following:

1. Preserve the tombs and their archeological contents
2. No excessive humidity ratio
3. No high air velocity near paintings

4. No mechanical vibrating installations

5. No permanent installations

6. Adequate uplighting

4.1 Historical Background

4.1.1 General

Tourism in Egypt is one of the national income supports, which helps increase the standard of living. The tombs of the kings, Luxor, Egypt, is considered one of the most important Egyptian pharaohs' heritages and one the favorite places for visiting. Human comfort is a critical issue to study to help more tourists to visit with no complaints. Relative humidity (moisture content), temperature, mechanical vibrations, lighting, noise, and insects are the main factors given attention, as they have significant influence on both humans and artifacts. The monuments of the Valley of the Kings are under serious threat, both from natural phenomena (flash flooding) and from the constant demands of tourism. The wall paintings in the valley represent the greatest collection of ancient art in the world. As Egypt's Ministry of Culture and Supreme Council of Antiquities are only too aware, dramatic steps must be and are being taken to control the situation, reverse current trends, and ensure the tombs' continuing survival. There are many factors, external and internal, that badly affect the artifacts.

4.1.1.1 External Factors

The biggest threat is floodwater penetration by flash flooding, as illustrated by the dramatic and tragic events during 1994. Incidents of heavy rains in the Theban Mountains are not unusual and have been noted from ancient times. Several tombs in the Valley of the Kings are completely choked or contain chambers that are thoroughly encumbered with the debris of flooding. In constructing schemes to prevent floodwater damage to tombs in the Valley of the Kings, one can learn lessons from the history of archeology. It is worthwhile to examine the context of those discoveries in which the contents of tombs were found dry and well preserved. Rainwater, windstorm, and other natural emergencies have a weak effect in the archeological artifacts but less impact than floodwater.

4.1.1.2 Internal Factors

We take these factors into consideration in this work to reach the optimum method for protection of the artifacts. Following are detailed representations of these threats and management guidelines.

4.1.1.2.1 Relative Humidity

There is a level of environmental moisture content (EMC) consistent with maximum chemical, physical, or biological stability. When the EMC is too low or too high, the associated relative humidity becomes a risk factor. Therefore, it is of prime importance to control the surrounding relative humidity within acceptable limits in order to minimize risk associated with moisture levels.

4.1.1.2.2 Temperature

Temperature is a very critical factor because of chemical changes that occur when temperature becomes too low or too high. Thermal energy not only accelerates aging, but also can magnify the effects of incorrect relative humidity. The American Society of Heating, Refrigerating, and Air-Conditioning Engineers (ASHRAE) has published recommended standards for thermal comfort parameters. Maintaining a building within the following ranges of temperature and relative humidity will satisfy the thermal comfort requirements of most occupants. It's very important to control room temperature to avoid bad effects.

4.1.1.2.3 Illumination Intensity

Overexposure to light can cause photochemical or photophysical changes for some materials, but it may be controlled by architecture, to design the lighting source out of exposure to artifacts.

4.1.1.2.4 Biological Attacks (Pests)

This includes some insect species, mold, fungi, and bacteria. Controlling of relative humidity and ventilation can control such attacks, and the fungi activity can be limited by reducing relative humidity levels. The fungi activity is demonstrated in Valley of the Kings; this activity has led to undesirable effects, such as shown in the tomb wall paintings in Figure 4.1. Most molds thrive at warmer temperatures. When combined with high levels of humidity (about 70% or higher), temperatures of 22 to 24°C can cause mold to develop, as shown in Table 4.1.

One of the prime motivators for this study was to be able to control the biological activity via ventilation to reduce the relative humidity levels.

FIGURE 4.1
Effect of moisture content in artifacts.

TABLE 4.1

Mold Temperature and Moisture Relationship from ASHRAE

	Optimum		Limits		
Species	°C	RH	°C	°C	RH
Aspergillus amstelodami	33	93%	10	42	71%
A. niger	33	>98%	12	43	78%
A. gumigatus	40	>97%	12	53	82%
Penicillium martensii	23	>98%	<5	32	79%
P. islandicum	31	>97%	10	38	83%
Stachybotrys atra	23	>8%	7	37	94%

Ventilation to reduce the relative humidity levels.

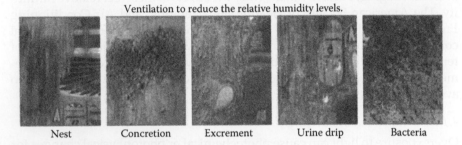

Nest   Concretion   Excrement   Urine drip   Bacteria

FIGURE 4.2
Typology and consequence of biological attacks (bat guano).

FIGURE 4.3
Effect of pest infestation in the wall painting.

The consequences from biological attacks and pest infestation can be described clearly, as shown in Figures 4.2 and 4.3.

4.1.1.2.5 Vibration

Vibration, which transmits from motors, compressors, tourist buses, seismic shocks, and the wind, can cause damage to sensitive objects. We should be more careful for the risk of vibration transfer through ductwork to works hung on adjacent walls or, in particular, active air drafts.

4.1.2 Main Factors Affecting Human Comfort

Human comfort is a very important issue to study; it is affected by

1. Temperature
2. Relative humidity
3. Local air speed

The above parameters can be controlled, and human comfort is also affected by health, age, activity, clothing, sex, food, location, season, etc. ASHRAE Standard 55-2010, *Thermal Environmental Conditions for Human Occupancy*, sets several principles that must be accomplished by the air distribution system. Furthermore, the study suggests different comfort conditions for Egyptian climate. The human comfort includes air quality, which influences the term *indoor air quality* (IAQ). This term concerns the air contaminants, which includes pollutants and overcrowding, tobacco smoke, and microbiological contamination. Any particles 10 microns or less are considered respirable. Generally, the smaller the particle, the greater the likelihood for penetration deep into the airways.

4.2 Background

4.2.1 General

Human comfort and indoor air quality (IAQ) in residential and commercial applications depend on many factors, including thermal regulation, control of internal and external sources of pollutants, supply of acceptable air, removal of unacceptable air, and occupants' activities. This part focuses on reviewing the previous thermal comfort experimental and numerical investigations carried out in enclosed spaces reported to provide an overall view of the research status at the commencement of the present work. The major experimental and numerical investigations of airflow characteristics in enclosed spaces are shown here in tabulated form. Comments and assessments follow such work.

4.2.2 Ventilation in Archeological Tombs of Valley of the Kings, Luxor

4.2.2.1 *Omar Abdel-Aziz and Khalil (2005)[1]*

This study, published in 2005, focused on the mechanical ventilation system's effect on airflow pattern, as well as temperature and relative humidity distribution inside three tombs, KV1, KV9, and KV62, to design an optimum heating, ventilation, and air conditioning (HVAC) airside system that provides comfort and air quality in the air-conditioned spaces. The study was carried out using computational fluid dynamics (CFD) simulation using a commercial CFD code that solved the continuity, momentum, energy, and species transport equations in addition to k-ε model equations for turbulence closure. The SIMPLEC algorithm was used for the pressure-velocity coupling, and a second-order upwind scheme was use for discretization of the governing equations. Mesh sizes used in the work exceeded 700,000 mesh volumes in one case, and all mesh sizes were above 100,000 mesh volumes. The outside air conditions, number of visitors, and airside system design effects on the tombs' airflow characteristics were studied in order to reach the optimum ventilation design.

4.2.2.1.1 *Rameses VII Tomb (KV1)*

The airflow distribution in its final steady pattern is a result of different interactions, such as the airside design, object distribution, thermal effects, occupancy movements, etc. The entrance zone is excluded from the tomb structure, as it doesn't represent airflow inside the enclosure. These volumes are discretized using the tetrahedral tool due to the complex geometry inherent in the model. Various mesh sizes could be obtained for each tomb. Different configurations are investigated:

- Either the left or right outlets are enabled, while the rest are treated as walls.
- The center floor-mounted outlets are enabled, while the rest are treated as walls.
- Both the left and right floor-mounted outlets are enabled, while the center outlets are treated as walls.

The type of ventilation used, mechanical supply or mechanical extraction or both, and the effect of the number of visitors on the airflow characteristics are considered. Visitors' bodies and faces are considered isothermal walls.

4.2.2.1.1.1 *General Flow Pattern*   The general flow pattern was studied via a velocity magnitude contour plot for a vertical mid-plane along the tomb axis (see Figure 4.4).

4.2.2.1.1.2 *Thermal Pattern*   Simulating February's maximum air design conditions showed a homogeneous temperature distribution throughout the tomb. On the other hand, in the case of simulating August's maximum air design conditions, there were larger temperature variations inside the tomb. Hence, the tombs should be close in cases of such harsh conditions in order to keep the tombs well preserved (see Figure 4.5).

It was found that the temperature gradient near the walls is almost similar in all cases. On the other hand, the mechanical supply ventilation system provided better thermal distribution in the occupation zone.

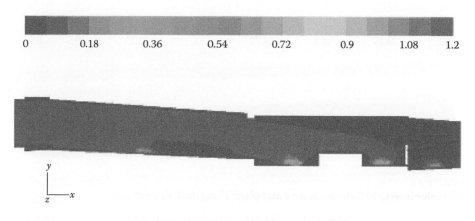

| 0 | 0.18 | 0.36 | 0.54 | 0.72 | 0.9 | 1.08 | 1.2 |

FIGURE 4.4
Velocity magnitude contours for a mid-plane along the KV1 tomb axis.

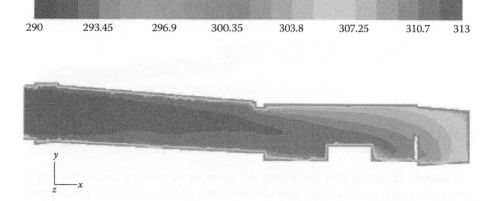

| 290 | 293.45 | 296.9 | 300.35 | 303.8 | 307.25 | 310.7 | 313 |

FIGURE 4.5
Temperature contours for a mid-plane along the KV1 tomb axis.

4.2.2.1.1.3 Relative Humidity Patterns   It was ensured that the higher the number of visitors inside the tomb, the larger the wall portion that was subjected to RH values higher than 60%. The results for no visitors inside the tomb showed higher RH values only due to the absence of internal heat loads, and hence lower airflow temperatures (see Figure 4.6).

4.2.2.1.2 Rameses V and Rameses VI Tomb (KV9)

The general flow pattern inside the KV9 tomb is affected by the airside system design. The extraction air outlets' location has a considerable effect on the main flow pattern; case 1 resulted in larger velocity magnitude until the first burial zone. Case 2 resulted in a similar situation, but the airflow velocity beyond the first burial zone was decreased (see Figure 4.7).

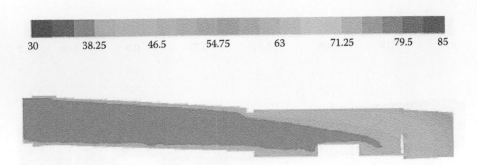

FIGURE 4.6
(See color insert.) RH% contours for a mid-plane along the KV1 tomb axis.

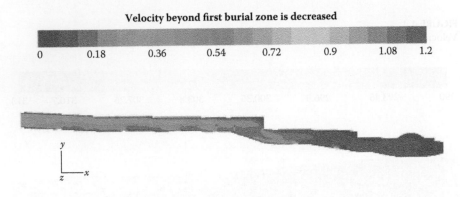

FIGURE 4.7
Velocity magnitude contours, m/s, for a mid-plane along the KV9 tomb axis.

It should be noted that the increased number of visitors inside the tomb resulted in increased velocities inside the tomb, as the visitors are considered obstacles for the airflow.

4.2.2.1.2.1 Relative Humidity Pattern The relative humidity distribution inside the KV9 tomb indicated the superiority of the mechanical extraction ventilation system over the mechanical supply, especially through centrally located air outlets (see Figure 4.8).

4.2.2.1.3 Tutankhamen Tomb (KV62)
The structure of KV62 is quite different from that of KV1 and KV9. Hence, the proposed airside system design relied mainly on fixed air grille locations. It is shown that if the mechanical supply system is used, higher air velocities are experienced near the walls, which are unacceptable, as they exceed the recommended velocity of 0.12 m/s. The temperature and relative humidity distribution presented ensure that the mechanical extraction ventilation system provides better distribution; however, this still results in excessive humidity inside the side rooms. Also, the effect of the number of visitors

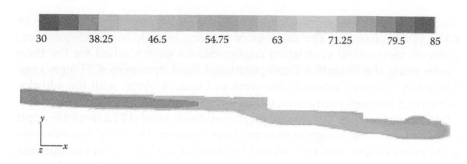

| 30 | 38.25 | 46.5 | 54.75 | 63 | 71.25 | 79.5 | 85 |

FIGURE 4.8
(See color insert.) RH% contours for a mid-plane along the KV9 tomb axis.

Thermal Pattern

| 298 | 300.25 | 302.5 | 304.75 | 307 | 309.25 | 311.5 | 313 |

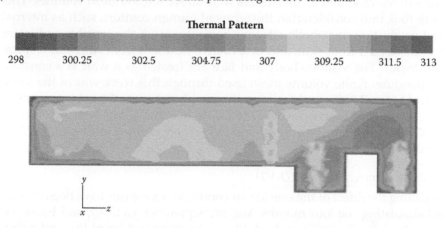

FIGURE 4.9
Temperature contours, K, for a transverse plane at $x = 18.5$ m (KV62).

was clearly identified, showing better performance with the lower number of visitors (17 visitors) (see Figure 4.9).

4.2.2.1.3.1 Thermal Pattern Finally, it was found that the main flow pattern of the free supplied air and floor-mounted extracts is slightly influenced by the extraction port locations. For each visitor group location, a corresponding proper airside design is suggested to provide the optimum utilization of the supplied air. The optimum utilization of the air movement to ventilate and reduce temperature can be attained by locating the extraction ports to minimize the recirculation zone and prevent the air short circuits. Ideally, the optimum airside design system can be attained, if the airflow is directed to pass all the enclosure areas before the extraction.

4.2.2.2 *Ezzeldin and Khalil (2006)*

This work investigates the human thermal comfort inside the three tombs KV1, KV9, and KV62 in Valley of the Kings, Luxor, for different proposed mechanical ventilation configurations. The study has been focused on KV62 (Tutankhamen). The number of visitors, outside air conditions, and different mechanical ventilation configurations were studied for the three tombs using the Fluent® 6.2 computational fluid dynamics (CFD) package. The study is a supplement to the work of Omar A. Aziz, with the addition of thermal comfort prediction, which is based on the predicted mean vote (PMV) model and the percentage predicted dissatisfied (PPD) model to apply these factors in the design of the ventilation system. The PMV equation contains many parameters that should be defined, such as the metabolic rate, mechanical work, clothes insulation, coefficients of heat transfer, and others, in order to be able to calculate mesh sizes exceeding 1,400,000 mesh volumes in one case, and all mesh sizes were above 100,000 mesh volumes. This work took into consideration the terms of human comfort, such as internal heat production (metabolic rate, mechanical power), the sensible and latent heat dissipated from skin, and evaporative heat loss from skin and respiratory losses. The visitors' body and face are treated as a wall at a constant temperature. Finite volume mesh used through this work was of the tetrahedral type. The thermal load equation is plugged into Fluent by means of the custom field function where the values of t_a and h_c are determined from the CFD results. Similarly, PMV and PPD where defined in Fluent and then calculated and displayed.

4.2.2.2.1 Rameses VII Tomb (KV1)

Depicting the effect of the outside air condition, four cases have been carried out simulating the four months August, September, October, and February, respectively, with different dry-bulb temperature and humidity and a constant number of visitors and mesh size. Kinetic energy and thermal pattern have been investigated also. The temperature contours for the cases depict the great effect the outside air condition has on the temperature patterns inside the tombs (see Figures 4.10 and 4.11).

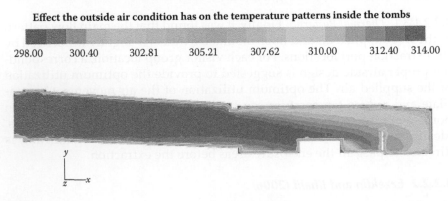

Effect the outside air condition has on the temperature patterns inside the tombs

298.00   300.40   302.81   305.21   307.62   310.00   312.40   314.00

FIGURE 4.10
(See color insert.) Temperature contours, K, for mid-plane, $z = 1.8$ m, KV1.

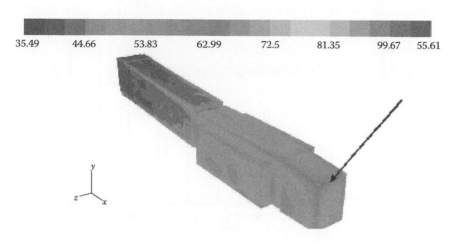

| 35.49 | 44.66 | 53.83 | 62.99 | 72.5 | 81.35 | 99.67 | 55.61 |

FIGURE 4.11
Relative humidity contours, %, at walls, KV1 (case 2).

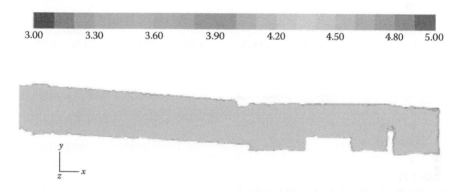

| 3.00 | 3.30 | 3.60 | 3.90 | 4.20 | 4.50 | 4.80 | 5.00 |

FIGURE 4.12
(**See color insert.**) PMV contours for a mid-plane at $z = 1.8$ m, KV1.

4.2.2.2.1.1 Relative Humidity The predictions of case 2 showed that the highest value of the relative humidity may reach 68%, which may initiate and support the growth of the fungi and mold on walls, ruin the paintings, and spoil these ancient colors. One should concentrate at the far end of the tomb, nearly at the dead end of the passage.

4.2.2.2.1.2 PMV and PPD Patterns The corresponding predictions of the PMV showed values in excess of +4, typically uncomfortable. The conditions are shown here to be unsatisfactory regarding comfort, as shown in Figures 4.12 and 4.13.

4.2.2.2.2 Rameses V and Rameses VI Tomb (KV9)

The construction of KV9 is a single-axis tomb, 116.84 m long and with a 4×4 m cross section with intermediate and end halls. Different parametric

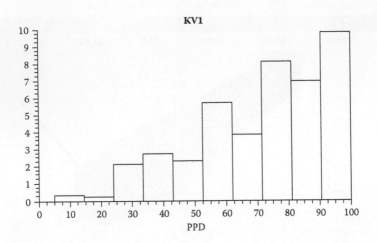

FIGURE 4.13
PPD histogram for KV1.

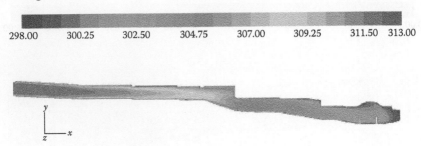

FIGURE 4.14
Temperature contours, K, for a middle plane, KV9.

case studies were investigated for this important tomb with different numbers of visitors to simulate possible situations of visitors' thermal and hygro loads. The contour plots in Figures 4.14 and 4.15 presented the thermal patterns and thermal comfort in the tomb. Predictions indicated that the mass flow rate has a significant effect of relative humidity all through the tomb. As for the increase of mass flow rate and outdoor relative humidity decrease, the indoor relative humidity reaches more than 80% near the end of the tomb passage.

4.2.2.2.3 Tutankhamen Tomb (KV62)

Many previous researches were reported for this tomb for its historical value. Numerical grids of different sizes were tested to obtain grid-independent results. Three different mesh sizes were used to simulate one case for the KV62 tomb; one grid of less than 300,000 tetrahedral mesh volume was used for preliminary investigation, while a final grid of more than 1,600,000 mesh volumes was used as a typical grid. The velocity near walls should not exceed 0.12 m/s[1] in order not to create any undesired effects on the paintings.

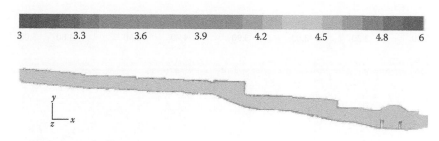

| 3 | 3.3 | 3.6 | 3.9 | 4.2 | 4.5 | 4.8 | 6 |

FIGURE 4.15
PMV contours for dissatisfied visitors, KV9.

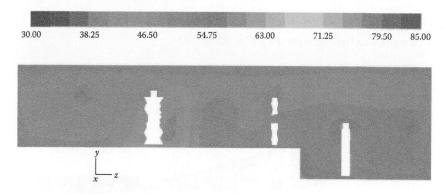

| 30.00 | 38.25 | 46.50 | 54.75 | 63.00 | 71.25 | 79.50 | 85.00 |

FIGURE 4.16
Relative humidity contours, %, for transverse plane $x = 17$ m, KV62.

The relatively high temperature gradient reaches 9°C near walls, as can be noticed in Figure 4.16, leading to severe thermal gradients.

The increase in the relative humidity has been noticed as the visitors' number increase reached 66%.

The penetration decreases as the visitors' number increases, as they function as an obstacle for the outside air, depicting the high-temperature gradient near walls, leading to high thermal gradients. Figure 4.17 demonstrates the predicted mean vote (PMV) in a section at $x = 17$ m in the tomb of King Tut (KV62). Figure 4.18 displays the predicted velocity contours in a transverse section of the tomb at $z = -1.5$ m.

4.3 Proposed Design Calculation Methodology

4.3.1 General

With the seriousness of the painted wall deterioration conditions, efforts were devoted to design a proper ventilation system for the various tombs

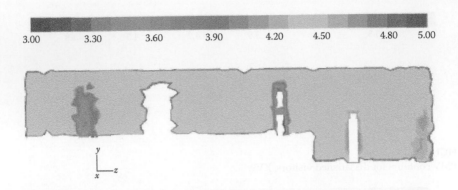

FIGURE 4.17
PMV contours for a transverse plane x = 17 m, KV62.

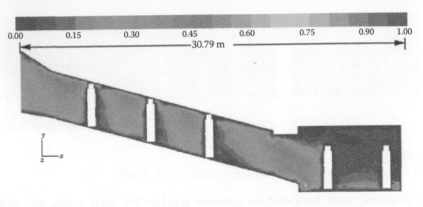

FIGURE 4.18
Velocity magnitude contours, m/s, for longitudinal plane z = –1.5 m, KV62.

and to investigate numerically airflow pattern, kinetic energy, and temperature and relative humidity distributions inside the archeological tombs of the Valley of the Kings, Thebes, to determine the effect of geometrical shape on the flow and relative humidity as well as comfort of visitors.

4.3.2 Classification of Tombs

The geometrical shape of the tomb according to number of passages can be classified into the following:

- Single (simple) passages
- Multiple (complicated) passages

In this study we chose a single-passage tomb with more than one room in the burial zone.
Four different tombs were selected according to a single passage.

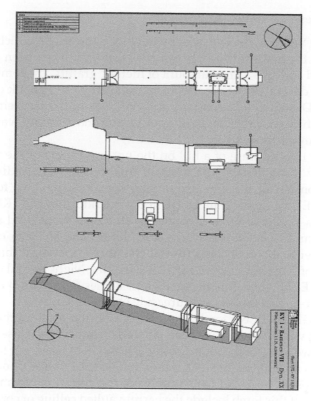

FIGURE 4.19
KV1 schematics. (From the Theban Mapping Project, http://www.thebanmappingproject.com.)

4.3.2.1 Rameses VII Tomb² (KV1)

Rameses VII's tomb, KV1, is located in the Valley of the Kings, Thebes. The entrance is cut into the base of a hill at the end of the first northwest branch wadi (valley). This unfinished tomb comprises an open entryway ramp (A), corridor (B), vaulted burial chamber (J), and unfinished chamber (K) with a rear recess, as shown in Figure 4.19. The walls are decorated with excerpts from the Book of the Gates (chamber K), Book of Caverns (corridor B), Book of the Earth (burial chamber J), Opening of the Mouth Ritual (corridor B), and the deceased with deities (corridor B, chamber K, burial chamber J); the ceilings are painted with motifs of flying vultures and astronomical figures. There are 135 Greek and several demotic, Coptic, and nineteenth-century graffiti in the tomb, indicating that KV1 has been accessible since antiquity. The KV1 tomb is considered a single straight-axis tomb oriented toward the northwest. The site is located at 25° 44′ N latitude and 32° 36′ E longitude. The tomb is 171.219 m above sea level. The maximum height is 4.25 m, the minimum width is 2.74 m, and the maximum width is 5.17 m. The tomb total length is 44.3 m. The floor area is 163.56 m², while the total volume is 463.01 m³.

4.3.2.2 Rameses IV Tomb² (KV2)

Found in the Valley of the Kings is the tomb of Rameses IV, which is located low down in the main valley, between KV7 and KV1. It has been open since antiquity and contains a large amount of hieratic graffiti. The tomb is mostly intact and is decorated with scenes from the Litany of Ra, Book of Caverns, Book of the Dead, Book of Amduat, and the Book of the Heavens. The sarcophagus is broken (probably in antiquity), and the mummy was relocated to the mummy cache in KV35.

KV2 is cut into the base of a hill on the northwest side of the main wadi of the Valley of the Kings, just south of the branch wadi leading to KV1. The tomb consists of three gently sloping corridors (B, C, D) followed by a chamber (E), a burial chamber (J), and a corridor beyond (K) with side chambers Ka–c. The tomb is decorated with scenes from the Litany of Ra (corridors B and C), Book of Caverns (corridors D and K), Book of the Dead (well chamber E), Book of Gates (burial chamber J), Imydwat (burial chamber J), Book of Nut (burial chamber J), Book of the Night (burial chamber J), Book of the Earth (gate Kb), deceased and deities (corridors B and K, side chambers Ka–c), and burial furniture (side chamber Kb).

The original plan of the tomb was altered after the death of the king, and the chamber, which would have been pillared chamber F, was used for burial chamber J. Two plans of the tomb are known: a plan of the whole tomb drawn on a papyrus now in the Turin Museum and a sketch of the doorway of the tomb on an ostracon found in the rubble at the entrance. Notable architectural features of this tomb include the barrel-vaulted ceiling of corridor D; the ramp through the floor of corridor D, gate E, and chamber E; the conversion of a pillared chamber into a burial chamber; and side chambers and recesses off the rear corridor K. Also unusual are the number of foundation deposit pits, although not all were used. Decoration unique to this tomb includes the representation of Shu and Nut from the Book of Nut on the ceiling of burial chamber J, the mummiform figures in Ka and Kc, and parts of the Book of Caverns, which appear for the first time in the Valley of the Kings. KV2 is one of the few tombs for which an ancient plan has survived. The tomb was frequently visited in antiquity, and graffiti are scattered throughout the tomb. In general, each visitor left his name, his profession, his origin, and personal comments about the tomb. There is a significant number of Coptic graffiti, including representations of saints and Coptic crosses. The tomb's dimensions are maximum height, 5.21 m; minimum width, 1.24 m; maximum width, 8.32 m; total length, 88.66 m; total area, 304.88 m²; and total volume, 1105.25 m³. Figure 4.20 illustrates the tomb of KV2.

4.3.2.3 Bay Tomb (KV13)²

Located in the Valley of the Kings in Egypt, the tomb of Bay was used for the burial of the noble Bay of the 19th Dynasty. It was later reused by

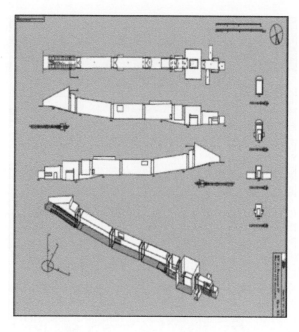

FIGURE 4.20
KV2 schematics. (From the Theban Mapping Project, http://www.thebanmappingproject.com.)

Amenherkhepshef and Mentuherkhepsef of the 20th Dynasty. The tomb of Bay is situated at the end of the southwest branch of the southwest wadi, close to the tombs of Seti II, Tausert, and Siptah. The architecture and decoration closely resemble that of the tomb of Queen Tausert. It consists of three corridors (B, C, D) followed by two chambers (E, F), two further corridors (G, H), two side chambers off the second (Ha–b), and a burial chamber (J). The tomb has suffered structural damage from floods, and all the ceilings of the tomb have collapsed. The walls were probably decorated originally with painted plaster and relief. Severe floods have caused the loss of the plaster, however, and now only traces of decoration remain in places where the artist was working on thin plaster and the chiselling cut into the bedrock. The remaining decoration echoes that in KV14 and represents the deceased with deities (corridors B and D) and parts of the Book of the Dead (corridor C). Of note, this is one of the rare nonroyal tombs cut in the valley during Dynasty 19. The tomb also demonstrates the late Rameside practice of reusing abandoned tombs for the burial of royal family members, containing two sarcophagi from this period. The tomb's dimensions are height, 2.53 m; width, 5.1 m; length, 5 m; area, 20.87 m²; and total volume, 54.23 m³ (Figure 4.21).

4.3.2.4 Seti II Tomb (KV15)²

KV15, the tomb of Seti II, has been known since antiquity and must have lay open during most of the classical period, judging from the 59 Greek and

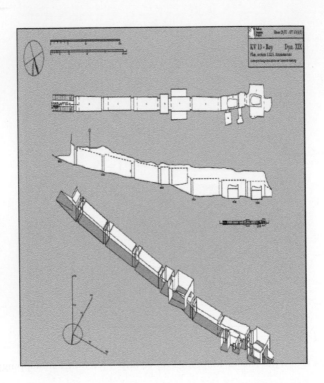

FIGURE 4.21
KV13 schematics. (From the Theban Mapping Project, http://www.thebanmappingproject.com.)

Latin graffiti found on its walls. The tomb was investigated superficially by Pococke, along with others who followed after him. However, it was Howard Carter who cleared most of the tomb between 1903 and 1904, though apparently the ritual well was never excavated. One may find the entrance to KV15, rather than having steps cut below a retaining wall, directly quarried into the base of an almost vertical cliff face at the head of the wadi running southwest from the main Valley of the Kings on the West Bank at Luxor (ancient Thebes). However, at present the tomb has been temporarily closed to allow the installation of new flooring, handrails, and lighting. It is expected to open soon. The history of the tomb is really unknown at this time. It is very likely that Seti II may have originally been buried with his wife, Tausret, in her tomb and later moved to this tomb, which appears to have been hastily and incompletely finished by Sethnakht (Setakht). In fact, the tomb may have originally been started for Seti II, but the work was interrupted at some point. This may have had to do with the reign of Amenmeses, if that king ruled concurrently with Seti II rather than before him. It appears that within the tomb, Seti's name was carved, erased, and then carved out once again. The erasure may be attributable to Amenmeses, or possibly Saptah. It has been suggested that Tausert then had her husband's name restored. The tomb, which takes a northwest-to-southeast axis, consists of a

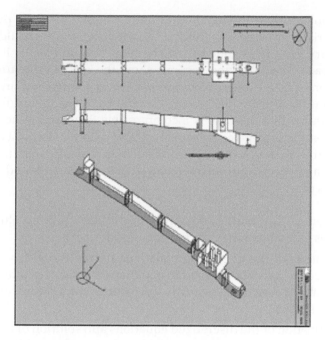

FIGURE 4.22
KV15 schematics. (From the Theban Mapping Project, http://www.thebanmappingproject.com.)

short entryway corridor followed by three long corridors, in turn followed by a well room. The well room then communicates with a four-pillared hall and then a makeshift burial chamber, formed from what would have been another corridor, where the king's sarcophagus was located. This corridor was hastily converted to a burial chamber at the time of Seti II's death. Its rough walls and ceiling were coated with plaster and decorated with paint. On the walls, the tomb's dimensions are height, 3.25 m; width, 2.77 m; length, 8.04 m; area, 21.51 m²; and total volume, 63.54 m³ (as shown in Figure 4.22). Anubis jackals on shrines and two rows of deities representing the followers of Ra and Osiris are placed over a lower register of prone mummiform figures on snake beds taken from the fifth division (p)/sixth hour (H). A figure of Nut with down-swept wings stretches along the length of the ceiling, and traces of what may be the Ba of Ra is painted above her head.

The present approach made use of Fluent 6.2 commercial codes. The numerical model included the following governing equations:

- Mass
- Momentum equations with the k-ε model or large eddy simulation (LES) as turbulence model closure
- Energy equation
- Species concentration equation

In order to construct the tetrahedral finite volume mesh the preprocessing was carried out using a Gambit® 2.2 mesh generator. It should be noted that the entry zone is excluded from the numerical model, as it is not an enclosed space. The following sections show the model structures for the different tombs. The numerical procedures incorporated within the computer program solver are described. It should be noted that in the present work we assume that the flow is incompressible flows, as in the airflow at low velocities; the segregated solver is typically used for simplicity. The governing integral equations for the conservation of mass, momentum, energy, turbulence, and species are solved using a control volume-based technique that consists of

- Division of the solution domain into discrete control volumes or meshes to establish the computational grid
- Integration of the governing equations on the individual control volumes to construct algebraic equations for the discrete dependent variables, such as velocities, pressure, temperature, and conserved scalars
- Linearization of the discretized equations and solution of the resultant linear equation system to yield updated values of the dependent variables

4.3.3 Mesh Generation

The mesh is generated using Gambit 2.2; using the GEOMETRY commands, the constructed lines are joined to form the faces, which are stitched together to form the computational volume. Air outlets are located on the raised floor in order to keep the archeological scheme unaltered. The air outlet details are shown in Figure 4.23. The air outlets may be located either near the side walls or at the floor center, allowing a diversity of airside system designs. Also, these outlets could be used for either mechanical supply or mechanical extraction of ventilation air.

The visitors are modeled as a composite volume formed of a rectangle representing the body and a cubic representing the face. Figure 4.24 shows the details and dimensions of a visitor model. These composite bodies are put

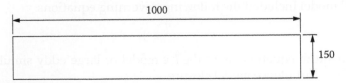

FIGURE 4.23
Floor-mounted air outlet details.

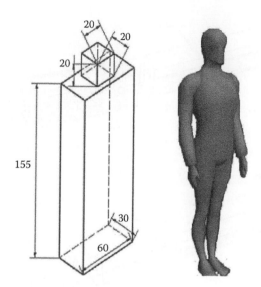

FIGURE 4.24
Visitor modeling assumption details.

in the most likely locations where the tomb paintings and decorations can be viewed.

The entrance zone of the tomb that is open to the atmosphere is excluded from the tomb structure, as it doesn't represent airflow inside the enclosure; Figures 4.25 to 4.28 represent the volume structure for KV1, KV2, KV13, and KV15, respectively. The air outlet locations can be clearly identified; however, the visitors are omitted from these figures in order to improve the visualization. These volumes are then discretized using the tetrahedral tool due to the complex geometry inherent in the model. Various mesh sizes could be obtained for each tomb.

4.3.4 Computational Design Studies

Detailed parametric investigations should be carried out in order to investigate the effects of various parameters on the tomb airflow and its characteristics. For each tomb, a set of parametric studies are investigated in order to suggest the proper design conditions. Grid dependence should be tested through CFD investigation in order to ensure that the grid proposed for the solution is of adequate size and that the results are grid independent. Air outlet location is a very important parameter in the airside system design, and hence different configurations are investigated for KV1, KV2, KV8, KV9, KV11, KV13, KV15, KV16, KV20, KV22, KV33+34, KV43, KV57, and KV62:

- The center outlets (in the middle centerline between the parallel walls) are enabled, while the rest are treated as walls.

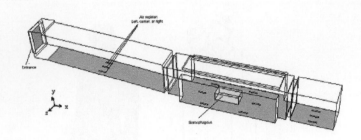

FIGURE 4.25
KV1 structure details.

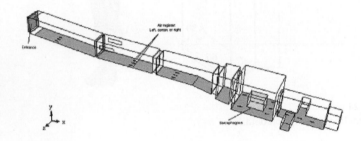

FIGURE 4.26
KV2 structure details.

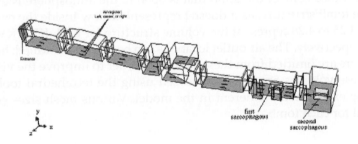

FIGURE 4.27
KV13 structure details.

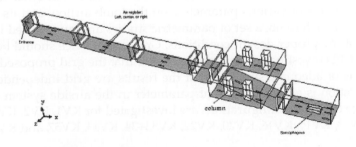

FIGURE 4.28
KV15 structure details.

- Both the left and right outlets are enabled, while the center outlets are treated as walls.
- The outside air conditions' effect on the CFD simulation of the tomb was also investigated in order to build sufficient knowledge of atmospheric conditions' effect on the airflow characteristics.

4.3.5 Boundary Conditions

4.3.5.1 Inlet Air Conditions

The inlet air conditions are taken as the average day maximum of 40°C (313 K) and 30% relative humidity (humidity ratio = 0.0138), representing August conditions. In addition, outside air conditions for September, October, and February are given. If the air is allowed to freely enter the tomb, the turbulence intensity could be assumed to be 6% and the length scale is assumed to be 1 m. However, if mechanical ventilation supply is incorporated, the length scale should be reduced to the smaller dimension of the air outlet, which is 0.15 m, while the turbulence intensity is kept constant at 6%. Furthermore, the flow is assumed to be normal to the inlet boundary.

4.3.5.2 Outlets

The air outlets are set as outflow conditions where the specification of the flow rate weighing can differ from one outlet to the other in order to allow more flexibility. More flow rate weighing is assigned to outlets near higher visitor population, whereas less flow rate weighing is assigned for outlets near the tomb entrance or near lower visitor population.

4.3.5.3 Walls

The walls are considered a slab, as they are deep inside the earth, and hence are considered to be at a constant temperature equal to the wet-bulb temperature of the outside air. Using the psychrometric chart, it can be found that the outside air wet-bulb temperature is 25°C. Also, it is assumed that the wall has zero species, water vapor, or diffusive flux. The no-slip condition is enabled for all walls, while using the standard wall function for near-wall treatment.

4.3.5.4 Visitors

The visitors' bodies are considered isothermal walls kept at the human skin temperature of 37°C due to the weak clothing of the tourist in Luxor. Furthermore, it is assumed that there is no diffusive flux. The visitors' faces are considered isothermal walls kept at the human skin temperature of 37°C as well. Also, it is assumed that there is a specified species mass fraction of

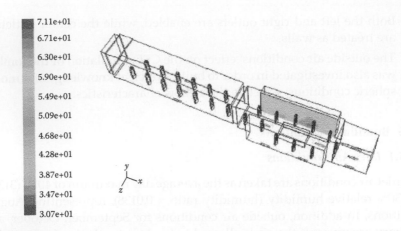

7.11e+01
6.71e+01
6.30e+01
5.90e+01
5.49e+01
5.09e+01
4.68e+01
4.28e+01
3.87e+01
3.47e+01
3.07e+01

FIGURE 4.29
(See color insert.) Relative humidity % near wall at KV1 with proposed design (24 visitors). (From Khalil, E. E., *Proceedings of ASHRAE RAL*, Athens, Greece, 2005.)

0.0411 kg$_\mathrm{w}$/kg$_\mathrm{d.a}$ in order to take into account the sweat effect in moisture gain to the tomb airflow.

Examples of the predicted flow pattern, path lines for the tomb of KV1, are shown in Figures 4.29 to 4.31 for a particular proposed number of visitors of 24 that may not adversely affect the tomb conditions. These results were obtained through a joint effort with Fluent experts.

An example of severely affected tombs is shown in Figure 4.32 for KV9, where condensation is predicted at the far end of the tomb; this matches the real situation presently at the tomb, which suffers deterioration of painting quality at this end.

4.4 Proposed Solution

The suitably addressed design should achieve a thorough investigation of the flow characteristics in tombs to design an adequate ventilation system that meets the requirements of artifact preservation.

4.4.1 Investigate Decay Mechanisms and Causes

The first aim is to study and analyze microclimate conditions of archeological indoor confined spaces, in order to identify the main causes of deterioration phenomena occurring. Field researches undertaken in the European area show that the most dangerous factors are related to the opening of these kinds of sites. On this basis, efforts should study and verify the suitability of these approaches in different cases in the contexts in the Valley of the Kings area,

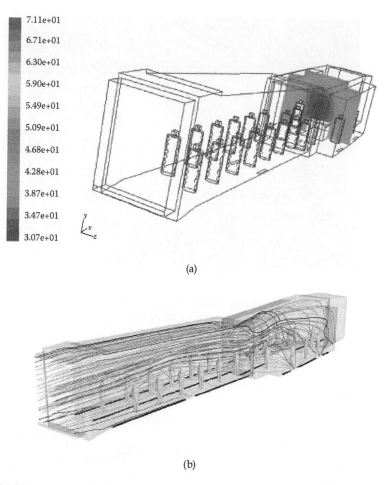

(a)

(b)

FIGURE 4.30

(a) Relative humidity at section at sarcophagus in KV1. (From Abdelaziz, O. O. A., and Khalil, E. E., CFD-Controlled Climate Design of the Archeological Tombs of Valley of the Kings, *Proceedings of CLIMAMED*, Second Mediterranean Congress of Climatization, Madrid, Spain, 2005.) (b) Path lines in Ramses VII KV1 tomb. (From Khalil, E. E., and Abdelaziz, O. O. A., *Fluent News*, 28, 2006.)

where not only environmental and climatic conditions are different, but also natural and social factors. Experimental multidisciplinary field campaigns will be carried out in both different climatic conditions and use of the site, in order to compare analysis results coming from a range of situations, which will vary from closed confined spaces and not accessible to visitors to open confined spaces with a high frequency of visitors. Where possible, the analysis and monitoring activities will be performed in couples of similar structures, characterized by opposite microclimate conditions—closed/open—in the same site, in order to study the respective responses to the impact of natural, environmental, and anthropogenic factors.

FIGURE 4.31
(See color insert.) Simulated relative humidity contours on wall artifacts in KV1. (From Khalil, E. E., and Abdelaziz, O. O. A., *Fluent News*, 28, 2006.)

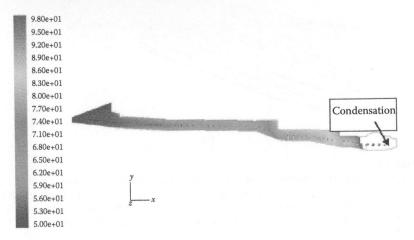

FIGURE 4.32
Contours of relative humidity, %, showing condensation at the end of KV9. At lower ventilation rate.

The measurement of the environmental parameters and their control (thermo-hygrometrical conditions that favor cycles of evaporation-condensation phenomena on the material, superficial condensation, deposition of pollutants, and formation of microorganisms), in order to evaluate the deteriorating phenomena, is the core of the analytical phase of the research. This analysis of environmental parameters will lead to the assessment of a behavior model of the thermal and hygrothermal cycles characterizing the interior microclimate, with the objective of measuring the tolerance threshold of the interior microclimate with respect to the impact of external factors.

Proposed field measurement campaigns should utilize instruments that are environmentally friendly, energy saving, sustainable, and have low impact for the sites. These instruments will be remote controlled, by means of a remote system able to transmit in real time the collected data, with the aim to reach a constant control of the interior and exterior environments. Recorded data for air temperature, relative humidity, CO, CO_2, volatile organic compound (VOC) concentrations, and air velocity should be taken 24 hours per day, all week long.

4.4.2 Design and Construction of Effective Ventilation System for Protection and Preservation

The ultimate achievable aim of the project is the elaboration of methodologies and procedures for a comprehensive approach to the conservation of archeological indoor confined spaces, taking into account the physical, chemical, and biological alterations, following a multidisciplinary approach with contributions coming from physics, geology, biology, archeology, and architecture. The results of multidisciplinary campaigns and monitoring activities on the different sites will provide a set of recommendations for their preservation, which will be based on real-time control, and conceived in order to provide to the actors responsible for the conservation of these sites and to stakeholders instruments useful for support decision of their protection.

The definition of new procedures for the preservation and protection of archeological indoor confined spaces will include the following:

- Design of a new robust ventilation system to ensure lower water vapor content in the air in the tombs.
- The identification/validation of the most suitable innovative procedures for environmental measurements (also modifying those existing in the market). These instruments will be environmentally friendly, energy saving, sustainable, remote controlled, and have low impact for the sites.
- The survey of the sites selected as case studies in the tombs of the Valley of the Kings area, the tests and monitoring activities to perform on them, and mapping of the environmental factors and risk areas on the microscale.
- The definition of a behavior model of thermic and hygrothermal cycles aiming to measure the tolerance threshold of the interior microclimate.
- The definition of a set of recommendations contributing to the identification and understanding of microclimate phenomena and suggesting compatible solutions.

Moreover, adding to the above standards and norms, the HVAC systems should be selected and designed based on the following considerations:

1. The climatic conditions.
2. Computability with architectural layout and aesthetics.
3. Overall economy of construction by establishing standard repetitive components.
4. Desirable interior environmental conditions.
5. Use of local material and equipment.
6. Durability and ease of maintenance.
7. The following codes and standards shall be used for HVAC mechanical systems design.

Figures 4.33 and 4.34 are two proposed ventilation systems that should be capable of providing adequate airflow rate controllable and linked to the local relative humidity measured at various locations in the tombs. The figures are for the simple KV1, where air is freely admitted from the entrance and is returned through the floor-mounted grilles connected by a raised floor flexible duct to extract fans outside the tomb. Figure 4.33 proposed simple extraction airflow to lower and control RH%.

The proposed simple control methodology is shown in Figure 4.34, and it includes the temperature, relative humidity, and IAQ sensors that would operate/stop/control the air extraction fan, as well as send messages to the site control to redirect the visitors to allow for time to lower the relative humidity inside the tomb.

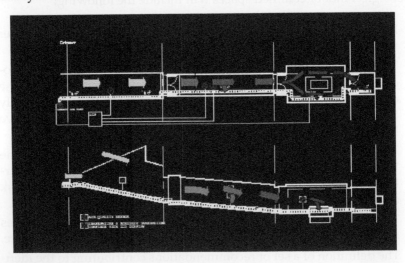

FIGURE 4.33
Proposed design of floor-mounted air extraction for KV1.

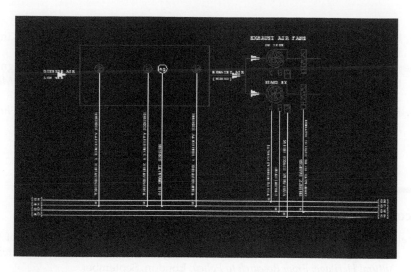

FIGURE 4.34
Proposed simple control methodology of air extraction to lower and control RH%.

4.5 Conclusions

The present computational procedure was taken a step further to investigate the airflow patterns, thermal behavior, and relative humidity in the tombs of the Valley of the Kings, a special case of enclosures. The aim here is to arrive at a comprehensive approach to the conservation of archeological indoor confined spaces, accounting for physical, chemical, and biological alterations. The results of multidisciplinary campaigns and monitoring activities on the different sites would provide a set of recommendations for their preservation, which will be based on real-time control, and conceived in order to provide to the actors responsible for the conservation of these sites and to stakeholders instruments useful to support decisions on their protection. The sample of results showed in this chapter calls for a new robust ventilation system to ensure lower water vapor content in the air in the tombs. The thermal behavior model of thermic and hygrothermal cycles in the tomb should aim at defining the tolerance threshold of the interior microclimate.

References

1. Abdelaziz, O. A. A., and Khalil, E. E. 2005. Predictions of air flow patterns and heat transfer in the tombs of the Valley of the Kings. Paper 358. *Proceedings of CLIMA*, October.

2. Theban Mapping Project. 2013. http://www.KV5.com.
3. Khalil, E. E. 2005. Indoor air climatic design of the tombs of Valley of Kings. In *Proceedings of ASHRAE RAL*, Athens, September.
4. Abdelaziz, O. A. A., and Khalil, E. E. 2005. CFD-controlled climate design of the archeological tombs of Valley of Kings. Paper 86. *Proceedings of CLIMAMED 2005*, 2nd Mediterranean Congress of Climatization, Madrid, February.
5. Khalil, E. E., and Abdelaziz, O. A. A. 2006. A dry passage to the afterlife. *Fluent News* 28.

Recommended Readings

Abdelaziz, O. A. A., and Khalil, E. E. 2004. CFD-controlled climate design of the archaeological tombs of the Valley of the Kings. *Proceedings of Sustaining Europe's Cultural Heritage: From Research to Policy*, London, September.

Khalil, E. E. 2004. Indoor air climatic design of the tombs of Valley of Kings. ROOMVENT 2004, Coimbra, Portugal, September.

Abdelaziz, O. A. A., and Khalil, E. E. 2004. CFD-controlled climate design of the archaeological tombs of Valley of Kings. *Proceedings of Indoor Climate of Buildings 2004*, Slovakia, November.

Abdelaziz, O. A. A., and Khalil, E. E. 2005. Air flow regimes and thermal patterns in climatized tombs in Valley of Kings. AIAA-2005-1444. January.

Abdelaziz, O. A. A., and Khalil, E. E. 2005. Indoor air flow regimes in the tombs of Valley of Kings. *Proceedings of International Conference on Energy and Environment, EE9*, Sharmelsheikh, March.

Abdelaziz, O. A. A., and Khalil, E. E. 2005. Modeling of indoor air quality and comfort in the tombs of Valley of Kings. Paper HT2005-72005. 2005 ASME Summer Heat Transfer Conference, San Francisco, CA, July.

Abdelaziz, O. A. A., and Khalil, E. E. 2005. Mathematical modeling of air flow and heat transfer-predictions of archeological tombs of the Valley of the Kings. Paper 185. *Proceedings of Indoor Air 2005*, China, September.

Khalil, E. E. 2005. Air flow regimes and thermal patterns in climatized tombs in Valley of Kings. Arab Construction World XXIII(6), September/October.

Abdelaziz, O. A. A., and Khalil, E. E. 2005. Understanding air flow patterns and thermal behaviour in "King TUT ANKH AMEN Tomb." Paper IMECE2005-80465. *Proceedings of IMECE 2005*, 2005 ASME International Mechanical Engineering Congress and Exposition, Orlando, FL, November.

Abdelaziz, O. A. A., and Khalil, E. E. 2006. LES versus k-ε turbulence modelling of air flow thermo physical characteristics in large underground archaeological facilities. AIAA Paper AIAA-2006-1105. January.

Khalil, E. E. 2006. 21st century CFD prediction of flow regimes and thermal patterns in 15th century BC tombs of the Valley of Kings. AIAA Paper AIAA-2006-0129. January.

Khalil, E. E. 2006. Controlled climate design of the archaeological tombs of Valley of the Kings. *Proceedings of Cold Climate 2006*, Moscow, May.

Ezzeldin, H. M., Mourad, S. M., and Khalil, E. E. 2006. Human thermal comfort in the tombs of the Valley of Kings using PMV model in non-air conditioned spaces. AIAA-2006-4171. *Proceedings of the 4th IECEC*, San Diego, June.

Abdelaziz, O. A. A., and Khalil, E. E. 2006. Air outlets locations effect on thermal and humidity patterns inside the archaeological tombs of the kings. *Proceedings of Healthy Buildings* IV, 221.

Ezzeldin, H. M., Mourad, S. M., and Khalil, E. E. 2006. Numerical simulation of thermal behaviour and human thermal comfort in the tombs of the Valley of Kings. *Proceedings of Healthy Buildings* V, 73.

Khalil, E. E., Abdelaziz, O. A. A., and El-Hariry, G. 2006. Climatic control inside the tombs of the Valley of Kings in Egypt. 2006 International Refrigeration and Air Conditioning Conference, Purdue, IN, July.

Abdelaziz, O. A. A., and Khalil, E. E. 2006. Proposed preservation index for ventilation system assessment in archaeological facilities. *Proceedings of Healthy Buildings* IV, 331.

Khalil, E. E. 2006. Air flow patterns and climatic control of the tombs of Valley of Kings. SET2006, 5th International Conference on Sustainable Energy Technologies, Vicenza, Italy, August 30–September 1.

Khalil, E. E. 2006. CFD, a tool for optimum airside system design inside archaeological tombs in the Valley of Kings. *Proceedings CLIMAMED 2006*, Lyon, November.

Khalil, E. E. 2006. Preserving the tombs of the pharaohs. *ASHRAE Journal* 34–38.

Khalil, E. E. 2007. Flow regimes and thermal patterns in 15th century BC tombs of the Valley of Kings. *Proceedings 2nd ME-EE Conference*, Greece, May.

Khalil, E. E. 2007. Air flow patterns and thermal behaviour in "King Tutankhamen Tomb." *International Review of Mechanical Engineering* 1(4), 444–450.

Khalil, E. E. 2007. Ventilation of the tombs of the Valley of Kings, Luxor and the Pyramid of Giza. *Proceedings IAQVEC 2007*, October.

Khalil, E. E. 2007. Air flow regimes and thermal patterns in the tombs in Valley of Kings. *Proceedings, 38th International Congress on Heating, Refrigerating and Air-Conditioning*, Belgrade, December. See also *KGH Journal* 37, 55–58, 2008 (in Serbian).

Khalil, E. E. 2008. CFD computations of flow regimes and thermal patterns in the tombs of the Valley of Kings. *Engineering Applications of Computational Fluid Mechanics* 2(1), 1–11.

Khalil, E. E. 2008. Air flow regimes and thermal patterns in the tombs in Valley of Kings. 10th Global Engineering Conference, Las Vegas, April.

Khalil, E. E. 2008. CFD applications for the preservation of the tombs of the Valley of Kings, Luxor. CHT-08-354. *Proceedings of CHT-08, ICHMT International Symposium on Advances in Computational Heat Transfer*, Marrakech, Morocco, May.

Khalil, E. E. 2009. Indoor air quality: CFD applications for the preservation of the tombs of the Valley of Kings, Luxor. In *Proceedings of the 4th IBPC*, Istanbul, June, pp. 767–772.

Khalil, E. E. 2009. CFD applications for the preservation of the tombs of the Valley of Kings, Luxor. Paper 127. *Proceedings of Building Simulation 2009*, Glasgow, July.

Khalil E. E. 2009. Energy efficient design of ventilation system for the preservation of the tombs of the Valley of Kings, Luxor. AIAA_2009_4573. IECEC, August.

Khalil, A. E. E., and Khalil, E. E. 2009. Air flow regimes and thermal patterns in climatized Church of Christ, Cairo. AIAA_2009_4574. IECEC, August.

Khalil, E. E. 2009. Ventilation system for the preservation of the tombs of the Valley of Kings, Luxor. Paper 55. *Proceedings of Healthy Buildings*, September.

Khalil, E. E. 2010. Thermal comfort and air quality in sustainable climate controlled healthcare applications. AIAA-2010-0802. AIAA, Orlando, FL, January.

5

Airflow in Places of Worship

The selection and control of the cooling, ventilation, and air conditioning systems to provide controlled internal conditions in historic and heavy-weighted buildings require knowledge not only of the effectiveness of the proposed system providing the required conditions quickly and efficiently, but also of the long-term effects of the system and any sudden fluctuations of the internal climatic conditions on the comfort and life of the building envelope itself and any works of art therein. Places of worship are of high spiritual nature and commonly architecturally elite. The occupants are usually praying either standing or standing, bowing, and kneeling. The level of activities should be accounted for in the thermal load calculations. Two examples are shown here: the Grand Mosque extension project in Mecca, Saudi Arabia, and St. Mary's Orthodox Church in Cairo, Egypt.

In the mosque, five obligatory prayers are performed daily; the average period of time for any prayer lasts from around 15 to 20 min. When performing a congregational prayer in mosques, Muslims are guided by a leader called imam. The imam stands at the front wall of the mosque facing mihrab, which is a semicircular niche in the wall of a mosque and indicates the qibla, that is, the direction of the Ka'ba in Mecca, and hence the direction that all Muslims should face when praying.

During prayer, the imam performs movements including postures such as standing, bowing, and sitting. Verses or chapters from the holy Qur'an are recited while standing. Therefore, for all intents and purposes the activities of all the participant prayers are synchronized behind the imam, and may be considered to be the same.

Therefore, within accepted Muslim practice, any intentional action resulting from unaccepted indoor thermal conditions or any movement departing from the actual performance of the prayer is not allowed and would make a person's prayer invalid. However, in Egypt and most other Islamic countries, participation in congregational prayer in mosques is very common. With some exceptions, all the mosques are normally occupied by males only, because females are highly encouraged to pray in their home, especially the five obligatory prayers.

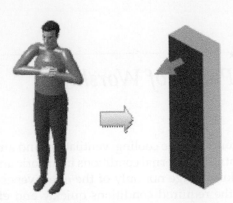

FIGURE 5.1
Worshiper modeling.

5.1 People and Comfort

5.1.1 Comfort Criteria

Thermal comfort is maintained when the body is in thermal equilibrium with its surroundings. The human body exchanges heat with the environment through convection, radiation, evaporation, and conduction to solid objects.

5.1.2 Effect of Air Movement

Air flow patterns have very important roles in transferring heat, species, and particulates into the space and have to be comprehensively investigated.

5.1.3 Acceptable Comfort Zone

The acceptable comfort zone shall be as prescribed by ASHRAE Standard 55.[2] The acceptable ranges of operative temperature and humidity for persons in a typical summer are 0.35 to 0.6 Clo and in winter 0.8 to 1.2 Clo at near sedentary (<1.2 met) activity levels. Considering comfort zone, air movement, and moisture content, in the present examples predictions were obtained by simulating a worshiper's body as a vertical rectangular box 1.75 m height × 0.25 m width × 0.5 m length, as shown in Figure 5.1.

5.2 Climatic Factors

5.2.1 Climatic Elements Affecting Natural Cooling

The local climate affects the building's energy efficiency, the comfort of its occupants, and its resistance to weathering.

5.2.2 Extrapolating Regional Weather Data to Specific Sites

The weather at the building site may differ from that at the weather station providing the climatological data used in the design.

5.2.2.1 Humidity

There is usually very little humidity data available from local second-order weather stations, and there are few generalizations that can be made about the amount of atmospheric moisture above a site, based on a description of the site's physical characteristics.

5.2.2.2 Solar Radiation

Solar radiation data are usually not available from local sources, either measured directly or extrapolated from cloud cover observation. It is possible to quantify the very local effects of site obstructions blocking solar radiation on-site hour by hour.

5.2.2.3 Temperature

Temperature varies geographically with elevation and surface type. Urban areas may have higher temperatures than the surrounding rural terrain. Temperature data are most commonly available from second-order local weather stations.

5.2.2.4 Wind

The most important climate data extrapolations occur with the wind. As with temperature, local records may be used to adjust the bin or hourly data. Local records of wind are, however, far less common than local temperature records, and are often of dubious accuracy due to poorly positioned or maintained instruments. Figure 5.2 shows an example of the ceiling diffusers at the Grand Mosque extension in Mecca, Saudi Arabia.[1]

5.3 Air Conditioning of Mosques: Ceiling Designs

A computational fluid dynamics (CFD) computer code is applied to predict the airflow pattern, thermal and humidity contours, and CO_2 concentrations at different planes and with different designs of air distribution systems, such as high-induction diffusers (HIDs), displacement flows, floor supply, low wall supply, etc. More than 10,000,000 grid nodes are used together with the most recent numerical modeling techniques and turbulence models. It is

FIGURE 5.2
Main Hall in Grand Mosque extension, Mecca, Saudi Arabia.

assumed that the mosque has no walls between the different zones. One internal zone is considered here; the y direction is the vertical measured from floor upward, and X and z are the other two orthogonal directions in the horizontal planes. The outdoor air temperature is assumed to be 318 K (45°C). The air velocity at the air outlet is assumed to be 1.6 m/s, resulting in a total of 12.8 air change per hour. The air is to flow from the supply grilles to the prayers and then to extract grilles with the aim to make the occupant comfortable with minimum energy expenditure. The question that is commonly asked is: Where would be the supply grilles' locations to enforce maximum comfort? The predicted flow patterns were obtained with this fine grid of 10 million nodes after a series of grid independency tests that were necessarily carried out. In mosques, usually Muslims pray standing, kneeling, and sitting. However, while standing, that usually takes a larger time portion; worshipers are at rest for 5 min reciting parts of the holy book. The following predictions were obtained for standing worshipers in line. Figure 5.3 represents the air velocity contours in a vertical X-Y plane normal to lines of prayers that may extend for 100 m in the Z direction as an example, for a plane in the north-south direction (perpendicular on rows of prayers).

Velocities were shown here not to exceed 0.2 m/s at occupancy levels to maintain comfort; the supply grilles are located in the ceiling, as also illustrated in Figure 5.2.

The corresponding air temperature contours are shown in Figure 5.4 at the same plane, and temperatures were of the order of 28°C for that air-cooled ventilation scheme. The height is about 6 m in most of the areas.

The focus is here on global airflow distribution, relative humidity, and carbon dioxide levels. Figure 5.5 was calculated by the author to demonstrate the relative humidity distribution in a typical vertical plane and in the

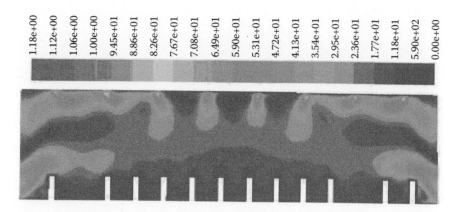

FIGURE 5.3
Velocity distribution at a vertical *x-y* plane, m/s.

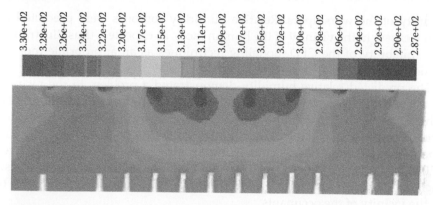

FIGURE 5.4
Temperature distributions at a vertical *x-y* plane, K°.

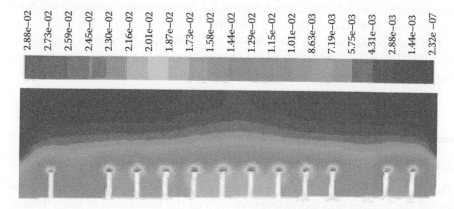

FIGURE 5.5
Percentage of relative humidity at a vertical *x-y* plane, RH%.

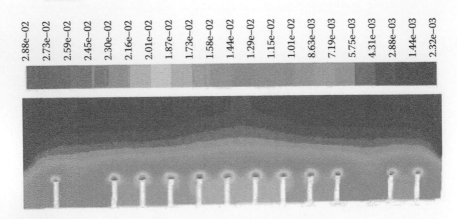

FIGURE 5.6
CO_2 concentration at vertical x-y plane, kg/kg.

vicinity of the worshipers. Relative humidity contours were predicted to show excess in the vicinity of the human faces, as expected, thus illustrating the suitability of the CFD modeling techniques as an appropriate tool for design support.

The predicted relative humidity demonstrated a problem in the vicinity of the worshipers that results from the excessive outdoor relative humidity, which can reach 70% in summer with temperatures of the order of 45°C. The corresponding cooling capacities are astronomical, and energy-efficient operating management was enforced to optimize the plant capacity of over 87,000 tons of refrigeration (more than 300,000 kW cooling). Predictions also included the carbon monoxide at breathing levels, as indicated in Figure 5.6, in the vicinity of the occupants.

It is also very important to understand the airflow characteristics at a plane around 1.8 m above the finished floor to realize the comfort at the worshipers' faces while standing. Figure 5.7 demonstrates the predicted flow patterns at a horizontal plane at $y = 1.8$ m.

The horizontal lines are lines of prayers; velocities are within the acceptable limits for places of worship. Another example of a smaller and simpler mosque will be demonstrated later in the chapter.

5.4 Air Conditioning of Churches: Ceiling Designs

The air distribution in archeological churches is very important for the comfort of the visitors and worshipers, as well as the well-being of the artifacts, building, and environment. The air conditioning of St. Mary's Church[3] in Cairo was one of the projects that the author was honored to be consulted

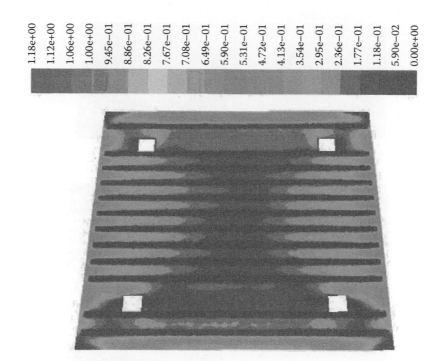

FIGURE 5.7
Velocity distributions (*x-z* horizontal plane) at $y = 1.8$ m.

on as a designer. Figure 5.8a and b shows some of the beautiful interior and exterior features of this church. In order to properly select the airflow parameters in such a pricey monument, computational techniques were utilized. The numerical model was used to investigate the airflow pattern, temperature, and relative humidity distributions inside the church main hall. The numerical model included the following governing equations:

- Viscous, k-ε model, or LES
- Energy
- Species

5.4.1 Model Architecture (Structure)

The church is located in Cairo; the main hall is modeled, as shown in Figure 5.8, as follows.

5.4.1.1 Inlet Air Conditions

The inlet air conditions are taken to be conditioned to the average day maximum of 40°C and 30% relative humidity, Egyptian code, representing August

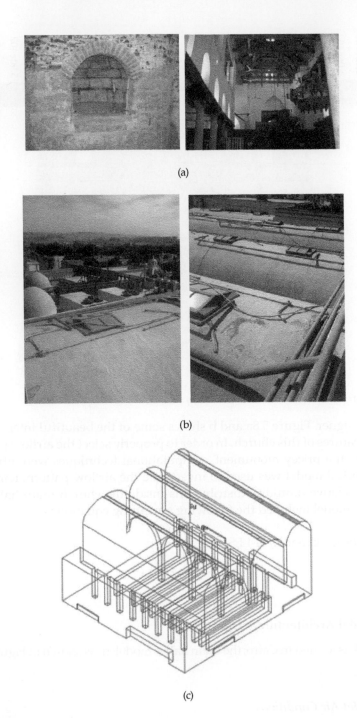

FIGURE 5.8
(a) Baby Christ hideaway (left) and St. Mary's Orthodox Church interior (right). (b) Exterior roof of St. Mary's Orthodox Church. (c) Isometric view for the church main hall.

conditions. The main hall is of 17.2 m × 18.2 m × height, which is variable with the dome's maximum of 9.3 m, a total volume of 2424 m³.

5.4.1.2 Air Inlets and Outlets

The air inlets are set as velocity inlet boundary conditions where velocity was set to 1.5625 m/s with a total of 12 air inlets, each of 0.4 m² area. This resulted in a total flow of 7 m³/s. The inlet air temperature was set to 287 K, with 0.008 water vapor mass fraction and negligible carbon dioxide concentration. The air change per hour (ACH) is chosen to be of 10. The air outlets are set as outflow conditions.

5.4.1.3 Walls

The walls are considered as a slab to have zero heat flux. The no-slip condition is enabled for all walls, while using the standard wall function for near-wall treatment.

5.4.1.4 Visitors' Bodies and Faces

The visitors' bodies are considered isothermal walls with a temperature of 310 K; the visitors' faces are considered isothermal walls kept at the human skin temperature of 310 K as well. Also it is assumed that there is a specified species mass fraction of 0.0411 $kg_w/kg_{d.a}$ in order to take into account the sweat effect in moisture gain. For carbon dioxide, a diffusive mass fraction of 0.0474 $kg_{co2}/kg_{d.a}$ is chosen.

The church hall model design incorporated 12 grilles for air introduction, each situated in between the ceiling arcs. The return grilles were situated near the ground. The model was used to simulate the situation during a prayer; consequently, the total number of visitors was set to 150 persons. More than 1,000,000 tetrahedral control volumes were used, and numerical convergence was better than 0.001%. Grid independence tests were performed with grids of 8,000,000, 900,000, and 1,000,000 tetrahedral volumes, and the grid used yielded grid-independent results to 2%. The model was used to simulate the situation during a prayer; consequently, the total number of visitors was set to 150 people. The total thermal load was 277 kW cooling, fresh air 1350 L/s. Loads from solar gain were 3 kW, roof thermal transmitted loads were 69 kW, while ventilation load was 79 kW.

5.4.2 Velocity Predictions

In the present archeological church the following are the velocity predictions for the model at hand in a y-z plane. Air was supplied from the ceiling openings, and the velocity contours were predicted as shown in Figure 5.9.

FIGURE 5.9
(**See color insert.**) Contours of velocity at $x = 4$ m.

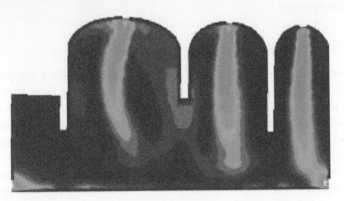

FIGURE 5.10
(**See color insert.**) Contours of velocity at $x = 15$ m.

Figure 5.10 shows the velocity contours at $x = 15$ m; near the other end of the church, the width of x varies between 0 and 17.2 m.

Figure 5.11 depicted the velocity contours in a transverse section at $z = 12.15$ m and indicated the prayers' standing locations. The velocities at these locations are well below 0.25 m/s, which ensured the disappearance of any drafts for the comfort of prayers and visitors. Figure 5.12 shows the velocity contours at a plane near the occupancy zone at 1.8 m above the finished floor.

The energy equation was solved to yield the temperature distribution at the various locations, taking into account the heat dissipated from the humans, equipment, and also the external heat sources in summer. Figure 5.13 indicated the temperature contours at a y-z plane at $x = 4$ m; temperatures are found to be homogeneously distributed and ensure comfort conditions.

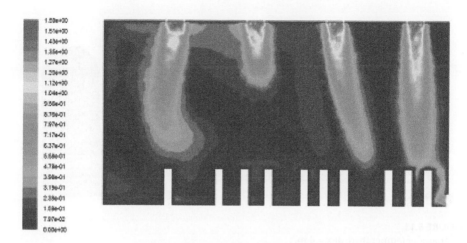

FIGURE 5.11
Contours of velocity at *z* = 12.15 m.

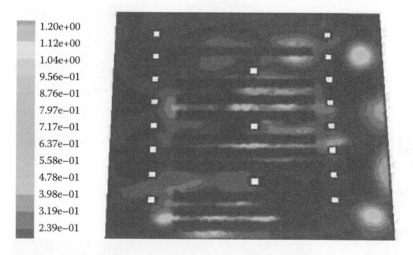

FIGURE 5.12
Contours of velocity at *y* = 1.8 m (people's faces).

Fanger's comfort model was used and PMV and PPD were predicted. PMV was around 0 ± 0.6 in the seating area of the prayers.

Figures 5.14 and 5.15 represent thermal patterns in transverse and horizontal planes where occupants are located; one can easily see temperatures of 30°C at the seating and standing locations. The remaining zones are at lower temperatures that can be as low as 17°C, bearing in mind that the on-coil temperatures leaving the ceiling supply grilles are typically 13°C. The temperature contours on a horizontal level at 1.8 m above the finished floor are shown in Figure 5.16.

FIGURE 5.13
Contours of temperature at $x = 4$ m.

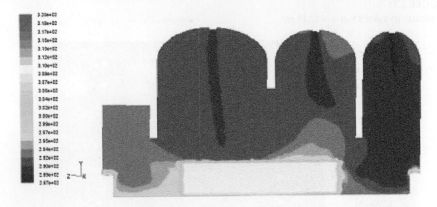

FIGURE 5.14
Contours of temperature at $x = 15$ m.

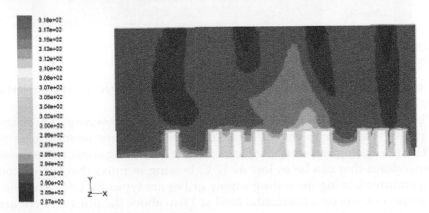

FIGURE 5.15
Contours of temperature at $z = 12.15$ m.

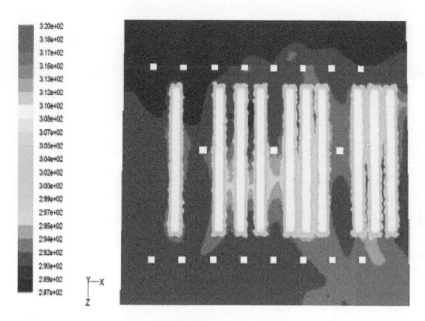

FIGURE 5.16
Contours of temperature at $y = 1.8$ m (people's faces).

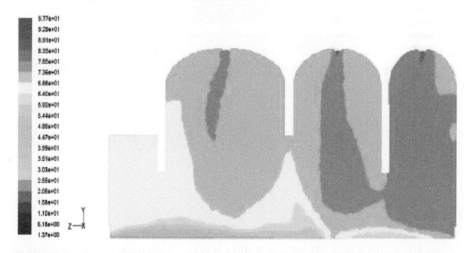

FIGURE 5.17
Contours of relative humidity at $x = 4$ m.

The relative humidity contours at various locations in the church are shown in Figures 5.17 and 5.18 at y-z at $x = 4$ and 15 m, respectively. The local values of RH% are in the vicinity of 50% at the occupancy level, as clearly indicated in the figures above, as the cooled supply air leaves the supply grilles at much higher values of 80% and more. Some disperse locations at near 1.8 m above the floor indicated high RH% values due to the presence

FIGURE 5.18
Contours of relative humidity at $x = 15$ m.

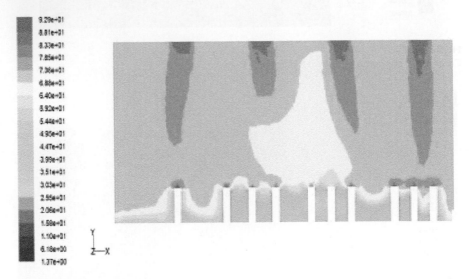

FIGURE 5.19
Contours of relative humidity at $z = 12.15$ m.

of candles and equipment. Figure 5.19 indicates some high values of relative humidity at the vicinity of the occupants' faces.

Local values of relative humidity percentage at the occupants' face level (1.8 m) are shown in Figure 5.20; excessive humidity percentages were observed in the rows of prayer level.

The carbon dioxide concentration contours at occupancy level in an x-z plane are shown in Figures 5.21 and 5.22 for consistency, indicating concentrations within the order of 0.0035. Predictions of carbon monoxide at a transverse plane at $z = 12.5$ m are shown in Figure 5.23.

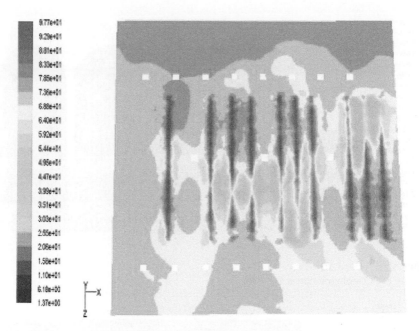

FIGURE 5.20
Contours of relative humidity at $y = 1.8$ m (people's faces).

FIGURE 5.21
Contours of CO mole fraction at $x = 4$ m.

Measurements of mean air temperature and relative humidity percentage were obtained with the aid of a hot-wire anemometer and electronic hygrometer with an accuracy of ±5%. These were compared to the corresponding predictions in Figures 5.24 and 5.25. Qualitative agreements were demonstrated with some discrepancies that are equally attributed to both

FIGURE 5.22
Contours of CO mole fraction at $x = 15$ m.

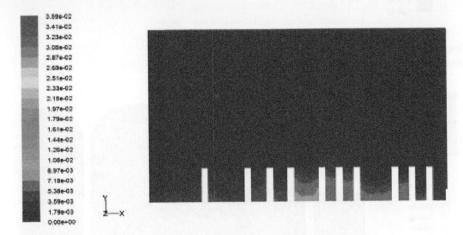

FIGURE 5.23
Contours of CO mole fraction kmol/m^3 at $z = 12.15$ m.

experimentations' accuracies and modeling assumptions. From the previous results, one can conclude that the airside designs have a strong influence on the relative humidity distribution, and consequently on the indoor air quality (IAQ). The location of the supply outlets plays a major role in this distribution. The extraction ports should be located in the right location to ensure comfort.

Due to the architectural design restrictions of archeological buildings such as this church, designers should perform this exercise to properly select the locations of supply and extract grilles in renovated systems in ancient buildings to yield better airflow, temperature, and relative humidity behavior.

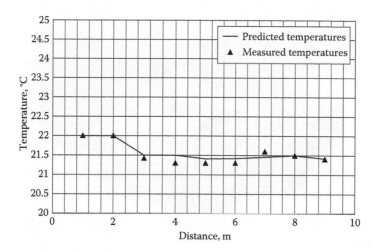

FIGURE 5.24
Measured and predicted air temperatures at 1.0 m above floor in the church. (From Khalil, E. E., Flow Regimes and Heat Transfer Patterns in Archeological Climatized Church of Christ, Cairo, IMECE-2012-85088, *Proceedings of ASME 2012 International Mechanical Engineering Congress and Exposition*, Houston, TX, 2012.)

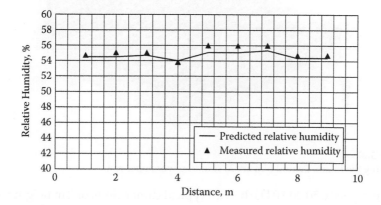

FIGURE 5.25
(**See color insert.**) Measured and predicted humidity percentage at 1.0 m above floor in the church. (From Khalil, E. E., Flow Regimes and Heat Transfer Patterns in Archeological Climatized Church of Christ, Cairo, IMECE-2012-85088, *Proceedings of ASME 2012 International Mechanical Engineering Congress and Exposition*, Houston, TX, 2012.)

5.5 Air Distribution in Mosques: Free-Stand Units

A more simple design of an archeological mosque (200 years old) was considered as another example. The mosque is located in Cairo, Egypt, and is air-conditioned with eight commercial floor (free) standing units, each of

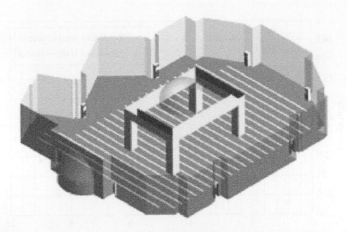

FIGURE 5.26
Mosque configuration with typical floor-mounted free-stand units.

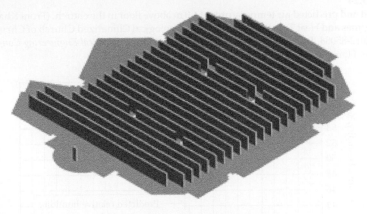

FIGURE 5.27
Simulating the imam. Behind him are 23 standing worshiper rows.

cooling capacity 50,000 BTU/h, with typical dimensions of 2 m height × 0.6 m length × 0.4 m width. The units are mounted on the floor and distributed about the perimeter of the mosque as shown in Figure 5.26.

The floor footprint is roughly 23 × 15 m, and estimating roughly space allocation per worshiper, the maximum number of worshipers within the mosque was found to be 912, including the imam. Similarly, based on previous assumptions for a worshiper body, a row of worshipers (standing shoulder to shoulder) is also modeled as a rectangular box with the same dimensions (height and width), while row length is as many times as a worshiper length, depending on how many worshipers are standing in the row. In other words, behind the imam are standing 911 worshipers arranged in 23 rows guided by their imam and following him by copying his ritual actions of worship, as shown in Figure 5.27.

TABLE 5.1

Boundaries and Occupant Data Details

Boundary	Value
Roof temperature	306 K
Wall temperature	305 K
Floor temperature	302 K
Outlet air temperature	286 K
Outlet air velocity	3.15 m/s
Initial air temperature	304 K

Occupant Data	Value
Gender	Male
Activity type	Standing, relaxed
Metabolic rate	1.2 met
Skin temperature	307 K
Clothing type	Light clothing
Clothing insulation	0.073 m²·K/w

Source: ASHRAE, *Fundamentals*, Atlanta, GA, 2009.

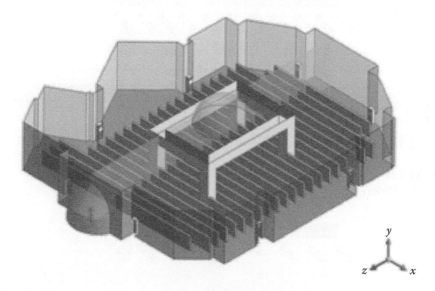

FIGURE 5.28
A vertical (*yz*) plane passing through the imam and all worshiper rows.

A thermal comfort model was used to represent the steady-state conditions. Thus, initial and boundary conditions are simply assumed according to moderate summer conditions in Cairo as shown in Table 5.1.

Results were plotted on an important vertical (*yz*) plane in the middle of the mosque, as shown in Figure 5.28.

The importance of this plane is due to

- Being as far away as possible from air conditioning units
- Passing through the imam and cutting the 23 following rows of worshipers

For simplicity, the results are arranged here such that the four predicted environmental parameters—air temperature, mean radiant temperature, air velocity, and relative humidity—are shown in Figures 5.29 to 5.32, respectively. The other dependent global comfort factors—predicted mean vote and predicted percentage dissatisfied—are shown in Figures 5.33 and 5.34.

The present predictions utilized the Fanger comfort model with its thermal sensation scales. PMV and PPD can be a reliable method to predict the performance of a heating, ventilation, and air conditioning (HVAC) system within a space in order to achieve the highest comfort level with minimum power consumption.

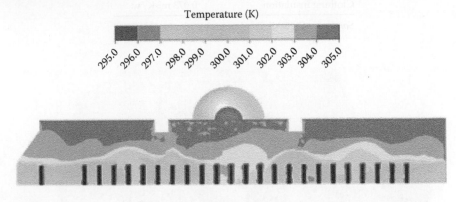

FIGURE 5.29
Predicted temperature contours (K).

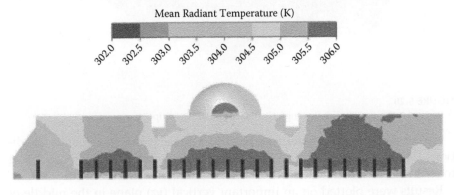

FIGURE 5.30
Predicted radiant temperature contours (K).

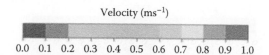

FIGURE 5.31
Predicted velocity contours (m/s).

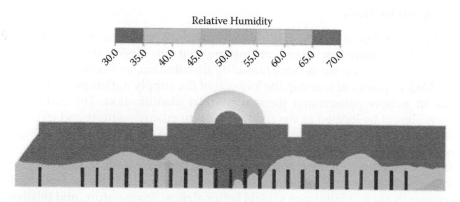

FIGURE 5.32
(See color insert.) Predicted relative humidity contours (%).

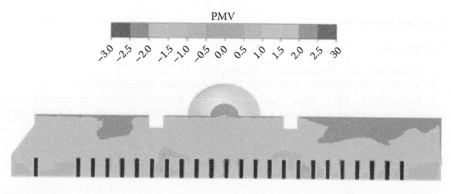

FIGURE 5.33
(See color insert.) Predicted mean vote contours.

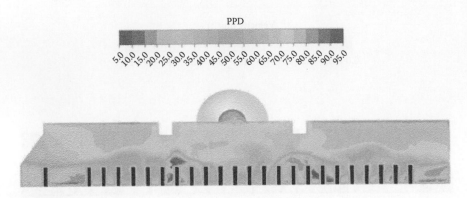

FIGURE 5.34
(See color insert.) Predicted percentage dissatisfied contours (%).

5.6 Conclusions

The computational airflow and thermal patterns predicted in this chapter dictated the more important conclusions that the airside designs have a strong influence on the relative humidity distribution, and consequently on the IAQ. In places of worship the location of the supply outlets plays a major role in airflow pattern and thermal comfort identification. The extraction ports should be located in the right location to ensure comfort and efficient utilization of the air conditioning energy. Due to the architectural design restrictions of archeological buildings and also places of worship, such as mosques and archeological churches, designers should perform this exercise to properly select the locations of supply and extract grilles in renovated systems in ancient buildings to yield better airflow, temperature, and relative humidity behavior.

References

1. ASHRAE. 2009. *Fundamentals*. Atlanta, GA.
2. ASHRAE Standard 55-2010. 2010. *Thermal comfort*. Atlanta, GA.
3. Khalil, E. E. 2012. Flow regimes and heat transfer patterns in archeological climatized Church of Christ, Cairo. IMECE-2012-85088. *Proceedings of ASME 2012 International Mechanical Engineering Congress and Exposition*, Houston, TX.

6

Airflow Patterns in Healthcare Facilities

This chapter is devoted to numerically investigating the optimum designs of various applications (devices and equipment) inside the different types of hospitals to obtain optimum energy utilization with the aid of computational fluid dynamics (CFD) techniques. Governing equations of mass, momentum, energy, and species are solved numerically to predict the airflow patterns and thermal behavior in operating theaters and healthcare facilities. This chapter is divided into several sections concerned with specific recommendations that should be followed for each application to achieve efficient energy utilization and indoor air quality.

This section is also devoted to the investigation of the influence of several airside designs on the efficiency of the flow of air-conditioned supply to create a sterile and comfort environment in surgical operating theaters. Energy efficiency improvement in air-conditioned healthcare applications was found to depend mainly on the application design configurations and operating parameters. The room airside design is one of the essential factors that strongly influence the heating, ventilation, and air conditioning (HVAC) airflow pattern, and consequently the air quality and comfort inside the special healthcare applications. Numerous examples of computed airflow patterns, moisture content distributions, and thermal images are demonstrated. This chapter introduces some recommendations for designs to facilitate the development of optimum energy-efficient design.

6.1 Airflow Characteristics for Comfort

This chapter is devoted to the investigation of the influence of the surgical operating theater architecture design on the efficiency of the flow of air-conditioned supply to create a sterile and comfort environment in the theater. It is also devoted to investigating the relationship between airflow movement and air age and the operating room architecture. The present work made use of a well-developed computational fluid dynamics (CFD) model based on a modified two-equation model k-ε that has been further developed and refined to predict the airflow regimes, turbulence characteristics, air temperature, and relative humidity distributions in a surgical operating theater. In the present work, the 3DHVAC program, previously developed and modified, and the

Fluent program are utilized to predict the air characteristics in surgical operating theaters. The airflow velocity and turbulent kinetic energy will be used as indicators of the indoor air quality (IAQ)-level performance of the present application. Actually, the central ceiling air supply perforated diffuser over the operating area would provide a solution for the complex situation of healthcare application by providing an acceptable IAQ level in the operating zone with enough energy efficiency. Energy efficiency in air-conditioned operating theaters was found to depend on the theater architecture design configurations and operating parameters. The room architecture design is one of the essential factors that strongly influence the HVAC airflow pattern, and consequently the air quality and comfort inside the surgical operating theaters. The present work also introduces some recommendations for architectural designs to facilitate the development of optimum HVAC systems.

Hospital operating theaters are classified as clean rooms. The HVAC system that services the operating theaters has a primary task: to limit the particulate concentration in the room and to provide the optimum conditions for hygiene and comfort. To build an HVAC system that is capable of fulfilling these requirements is a great challenge for designers. The teams that work in the operating theater (surgeons, anesthetists, nurses, and orderlies) interact with the room environment. Most infection possibilities are ascribed to respiratory diseases. The exposure to anesthetic gases is the most common problem that faces the surgery staff and even the patients. The critical factor for the air quality is the efficacy of the air conditioning and filtration system, and this efficacy depends on many design aspects, not only the airflow characteristics. The airflow distribution pattern plays an important role to reduce the surgery staff and patients' exposure to the hazards. Good design of HVAC systems creates this optimum flow pattern, and consequently the required protective area.

6.2 Early Practice of Airflow Design

There are several recommendations about the optimum airflow design in critical areas such as surgical operating theaters; however, there are few publications on this topic. These recommendations succeeded the laminar-linear airflow concept, which was introduced to provide restricted air movement. The directional and free turbulence airflow provides a clean environment. Most of the guidelines advise the use of vertically downward flow and don't recommend any horizontal supply air configurations. Supply airflow from ceiling- or high wall-mounted and extracting air from several ports located near the floor is proposed to maintain the downward flow. There are some considerations to avoid direct airflow toward the patients and to provide a sufficient air stream to the surgery staff in the operating theaters. Indeed, these guidelines are sufficient in the present status, but the problem that faces the HVAC system designers is how to build a system that is capable of following these guidelines efficiently.

6.3 Present Observations

Most of the surgical operating theater designs suffer from improper airflow distributions that increase the infection possibilities among the surgery staff and patients. Such a situation leads us to identify the recommended and not recommended airflow patterns in operating theaters. The first rule that should be followed is creating a protective area around the patients and surgery team, or at least at the respiration level of them, as shown in Figure 6.1. The intensity of pollution is relative to the location in the operating theaters. This intensity increases nearer to and in the vicinity of the operating area or the extract ports, and decreases nearer to the clean air supply outlets. It is therefore proposed to divide the theater space into several zones and analyze the nature of the flow in each zone and the interaction between the different zones. Therefore, the whole domain can be divided to three main zones: supply zone, activity zone, and dirty or exhaust zone, as shown in Figure 6.1. The activity zone lies between 1 and 2 m height to contain the patient and surgical staff. From

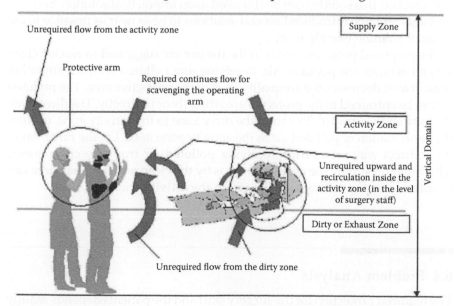

FIGURE 6.1
(See color insert.) Airflow nature in the surgical operating theaters.

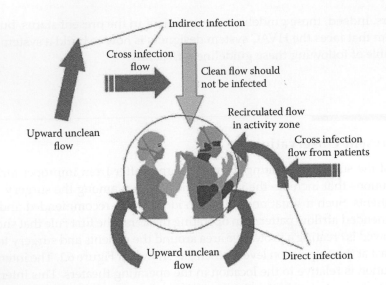

FIGURE 6.2
Infection sources of the protective area in the activity zone.

the practical viewpoint, interaction among the three zones in the forward and reverse directions exists. Only the complete "piston effect" flow can prevent backward or reverse interaction from the pollutant zones to less pollutant or clean zones.[7]

In practice, this solution can be classified as an impractical solution. So, most previous research introduced several solutions to be as near as possible to the ultimate solution (see Figure 6.2).

The proposed protective areas in the theater are suggested to receive clean air and exhaust the pollutant air. Receiving any polluted air from any other zone or area decreases the air quality level of the protective area. The polluted air can be entrained to the protected area directly or indirectly. The direct way stems from the upward flow from the dirty zone to the activity zone, or from the recirculation of polluted air in the activity zone itself. On the other hand, the indirect way arises from receiving polluted air from the clean zones, which should supply the protective areas by the clean air. The clean area can be infected by the recirculated flow from the activity and dirty zones.

6.4 Problem Analysis

The required protection for the surgery staff and the patients depends mainly on the airflow distribution pattern. Proper design of the HVAC airside would reduce the risks of infection with energy and comfortable conditions gains.

Several attempts were carried out to simulate the airflow distribution pattern, experimentally and numerically, to get the proper HVAC airside design, as indicated earlier by Kameel,[1] Kameel and Khalil,[2] Liu and Moser,[3] and Khalil.[4,5] All previous attempts found that air outlets and extracts are critical parameters responsible for obtaining a contaminant-free environment.

From the present observations, one can conclude that the complete laminar vertically downward flow can prevent any upward or recirculated flow from the dirty or activity zones. Using the complete piston displacement effect design can attain this. This design can be realized using a complete perforated ceiling as the supply of the operating theater in the present study. Moreover, it is supposed that to reach the complete laminar downward flow and the sterile environment, the extract air will be extracted from the floor to prevent any airflow recirculation.

The proper tactical airflow distribution is required in all applications in the surgical operating rooms to ensure comfort of occupants. The airflow distribution in its final steady pattern is a result of different interactions, such as the airside design, object distribution, thermal effects, occupancy movements, etc.[2,4,5] The present analyses utilized the numerical schemes embedded in 3DHVAC.[1] The governing equations of mass, three momentum, energy, and species concentrations are solved at an orthogonal three-dimensional grid arrangement that maps the flow field in the operating room.

6.5 Modeled Equations

The present computer program solves the differential equations governing the transport of mass, three momentum components, and energy in 3D configurations under steady conditions. The different governing partial differential equations are typically expressed in a general form as

$$\frac{\partial}{\partial x}\rho U\Phi + \frac{\partial}{\partial y}\rho V\Phi + \frac{\partial}{\partial z}\rho W\Phi$$

$$= \frac{\partial}{\partial x}\left(\Gamma_{\Phi,\mathit{eff}}\frac{\partial\Phi}{\partial x}\right) + \frac{\partial}{\partial y}\left(\Gamma_{\Phi,\mathit{eff}}\frac{\partial\Phi}{\partial y}\right) + \frac{\partial}{\partial z}\left(\Gamma_{\Phi,\mathit{eff}}\frac{\partial\Phi}{\partial z}\right) + S_{\Phi} \qquad (6.1)$$

where

ρ	= air density, kg/m^3
Φ	= dependent variable
$S\Phi$	= source term of Φ
U, V, W	= velocity vector
$\Gamma_{\Phi,\mathit{eff}}$	= effective diffusion coefficient

TABLE 6.1

Terms of Partial Differential Equations (PDEs), Equation 6.1

	$\wp$	$\wp_{\wp,eff}$	$S_\wp$
Continuity	1	0	0
x momentum	U	μ_{eff}	$-\partial P/\partial x + \rho g_x$
y momentum	V	μ_{eff}	$-\partial P/\partial y + \rho g_y(1 + \beta \Delta t)$
z momentum	W	μ_{eff}	$-\partial P/\partial z + \rho g_z$
H equation	H	μ_{eff}/σ_H	S_H
k equation	k	μ_{eff}/σ_k	$G - \rho\,\varepsilon$
ε equation	ε	$\mu_{eff}/\sigma_\varepsilon$	$C_1\,\varepsilon\,G/k - C_2\,\rho\,\varepsilon^2/k$

$\mu_{eff} = \mu_{lam} + \mu_t$ $\mu_t = \rho\,C\mu\,k^2/\varepsilon$

$G = \mu t\,[2\{(\partial U/\partial x)^2 + (\partial V/\partial y)^2 + (\partial W/\partial z)^2\} + (\partial U/\partial y + \partial V/\partial x)^2 +$
$(\partial V/\partial z + \partial W/\partial y)^2 + (\partial U/\partial z + \partial W/\partial x)^2]$

$C_1 = 1.44, C_2 = 1.92, C_\mu = 0.09$

$\sigma_H = 0.9, \sigma_{RH} = 0.9, \sigma_\tau = 0.9, \sigma_k = 0.9, \sigma_\varepsilon = 1.225$

The effective diffusion coefficients and source terms for the various differential equations are listed in Table 6.1. Details of the modeling technique and assumptions can be found in references 1, 2, and 4.

Table 6.2 represents different types of these trials. Some of these solutions already exist in real practice, and the others are virtual and are proposed for future implementation.

Table 6.2 introduces an analysis for each design based on the computational results obtained solving Equation 6.1 at 800,000 to 1,000,000 cells. This analysis of airside designs and previous research of energy efficiency and IAQ performance indicated the necessity of careful selection of the optimum locations of the supply outlets and extraction ports. The cross- or horizontal flow was found to result in what is termed inefficient sick design, and should be excluded as an airside design in the future. On the other hand, the downward flow that is developed by separate individual ceiling supply diffusers doesn't adequately furnish the functionality of the downward flow for providing a sterile environment in the operating area. Indeed, the relocation of separate diffusers over the operating area could lead to sufficient downward flow without any upward flow over the operating area.

Actually, the central ceiling supply plenum and perforated diffuser over the operating area provides a solution for the complex situation of healthcare applications by providing an acceptable IAQ level in the operating zone with proper comfort and flow pattern. The partial walls are recommended as a good design parameter to provide directed flow and protect the supplied clean airflow from the cross-horizontal infection. The upper level of extraction participates in the flow, guiding and preventing any upward flow from mixing with the clean air.

TABLE 6.2

Different Types of Airside Designs of the HVAC Systems in Surgical Operating Theaters

Supply Designs			
	 Horizontal Cross-Flow	 Distributed Partially Ceiling Flow	Centralized Partially Ceiling Flow
	Airflow can be supplied from two opposite walls or from the four walls; also can be supplied from corners.	Airflow is supplied from multiperforated ceiling diffusers distributed in the ceiling, especially around the surgery zone.	Airflow is supplied from multiperforated ceiling diffusers distributed in the ceiling center.
Extract Port Designs			
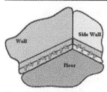 One level and near floor extract around the room perimeter.	The horizontal cross-flow is not recommended with all types of extracts due to the strong horizontal jet flow interactions and consequent effects on the airflow pattern and turbulence in the operating area and the activity zone, which enhances the infection. The infection possibility is increased with this type of supply air arrangements. This type is not energy efficient due to the presence of air short circuits between the supply and the extraction due to the nearing of the outlets and inlets.	The presence of the upward airflow in the operating area does not provide optimum protection to the patients or surgery team. The strength of the side recirculation zones near the walls and ceiling is increased due to the exclusion of the upper extract ports.	This design results in increasing the strength of the recirculation zones near the walls and ceiling. It provides complete protection of the operating area. Using the partial wall is highly recommended.
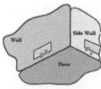 One level and near floor extract from the four walls or four corners.			
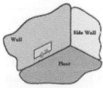 One level and near floor extract from the two opposite walls.			This design is not recommended due to the presence of large areas of recirculation zones.

Continued

TABLE 6.2 (*Continued*)

Different Types of Airside Designs of the HVAC Systems in Surgical Operating Theaters

Supply Designs			
	Horizontal Cross-Flow	Distributed Partially Ceiling Flow	Centralized Partially Ceiling Flow
	Airflow can be supplied from two opposite walls or from the four walls; also can be supplied from corners.	Airflow is supplied from multiperforated ceiling diffusers distributed in the ceiling, especially around the surgery zone.	Airflow is supplied from multiperforated ceiling diffusers distributed in the ceiling center.

Extract Port Designs			
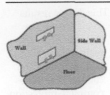 Two levels extract from two opposite walls.	Not applicable	The upper extraction level participates to decrease the recirculation.	This design is recommended. The extract ports can be located on the wider opposite walls to enhance the design efficacy.
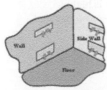 Two levels extract from four walls or four corners.	Not applicable		This design is highly recommended to provide a complete clean environment with fewer areas of recirculation.

6.6 Flow Pattern Analyses

The present section is devoted to analyzing the performance of HVAC system design in a practical full-scale operating theater with 1200 beds in an Egyptian modern teaching hospital, New Kasr El-Aini Teaching Hospital, Cairo University.[1] Figure 6.3 shows the operating theater configuration. The experimental program comprised field measurements of air velocity, temperature, and relative humidity, in the vicinity of supply air diffusers and the operating area. Comparisons between the measured and predicted flow patterns are shown and analyzed. In order to validate the 3DHVAC program with a real air-conditioned application, an extensive experimental program

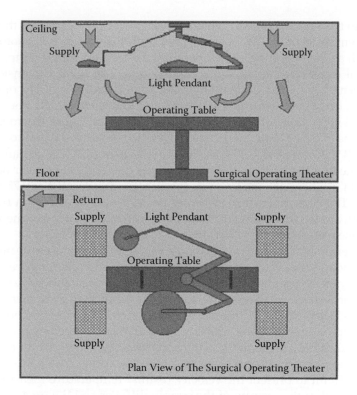

FIGURE 6.3
Schematic layout of the surgical operating theater (elevation and plan views).

was carried out to measure velocity, temperature, and relative humidity profiles. The experimental results illustrate the distinct interaction of the HVAC system design parameters. The present work also introduced some recommendations to ensure prevailing aseptic conditions.

6.6.1 Experimental Facility

The HVAC system that serves the hospital under consideration is an all-air system (variable air volume (VAV)). The HVAC system is in operation most of the year, and it supplies the different zones in the hospital by air that satisfies the healthcare recommendations according to ASHRAE standards.[6] The supplied airflow to the operating theaters was designed through four square ceiling perforated diffusers, each 0.6 × 0.6 m; the air was supplied through a high-efficiency particulate air filter (HEPA). The four supplying diffusers in each surgical operating theater have the same relative dimensions in each theater; however, there are differences in the sizes of each theater. Figure 6.3 represents a schematic layout of typical surgical operating theaters in a teaching hospital; the figure represents elevation and plan views of the surgical operating theater. It also indicates the positions of the supply air outlets relative to the operating table, which was located in the center of the room. The present experiments were performed with the presence of room furniture, such as the operating table, operating devices, operating lamp, and all additional accessories, as shown in Figure 6.3a. The room dimensions are 6.6 m in length (L), 4.0 m in width (w), and 3.0 m in height (H). The operating table has a length of 2.0 m and a width of 0.5 m and is 1.0 m high. Four square perforated supply air diffusers were located at the ceiling of the room with dimensions of 0.6 × 0.6 m; their centers were located at x, y equal to (1.7, 1.3), (4.9, 1.3), (4.9, 2.7), and (1.7, 2.7), as shown in Figure 6.3a. The exhaust ports were located on the left wall with dimensions of 0.5 × 0.3 m (lower one) and 0.5 × 0.2 m (higher one); the lower port center point is located at 0.75 m from the floor. The higher return port center point is located at 2.25 m from the floor. Both are located at lateral direction y = 3.6 m, with z being the vertical location measured from the floor up to the ceiling at z = H = 3 m.

The experimental procedure of Khalil and Kameel[7] introduces the measurements of downward velocity component (W m/s) and temperature (t °C) in the vicinity of the supply diffuser. The present measurements were obtained under steady-state conditions. The supply air conditions were measured as 0.31 m/s velocity and 19.5°C; for more details on the experimental procedure, see the original publications. Velocity measurements were performed with the aid of thermal anemometry, which consists of a probe (1 mm diameter) with an accuracy of 5%. The measuring probe is associated with an electronic feedback circuit to present the velocity results. Temperature measurements were obtained in the vicinity of the supply diffuser and the operating table with the aid of a thermocouple with an accuracy of 2%. The measurement errors were analyzed and the velocity measurements were found accurate within 5%, and temperature measurements were found accurate within 2%.

6.6.2 Numerical Program

The 3DHVAC program is a full three-dimensional flow solver that is based on the SIMPLE algorithm that solves finite difference equations governing the transport of mass, three momentum, energy, species, and turbulence entities. The present version of 3DHVAC program was produced by new modifications in the grid generation method 800,000 grid nodes, which were based on the new proposed hyperbolic grid generation formulas. The additional grid nodes and the using of new grid generation formulas will share in the modification of the model's ability to predict flow characteristics. Also, the partial differential equations for solving the relative humidity and contaminant age are introduced with required modifications in the program. More validation tests were also introduced to verify the ability of the program to obtain airflow characteristics.

Khalil[8] developed a computer program to predict the airflow velocities by solving the finite difference form of the governing equations using the SIMPLE numerical algorithm (Semi-Implicit Method for Pressure-Linked

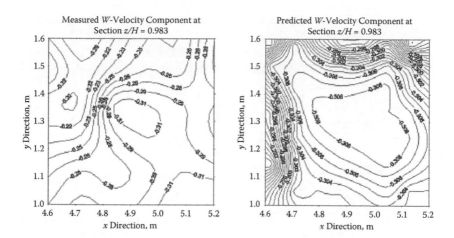

FIGURE 6.4

Comparisons between measured and predicted downward velocity components at horizontal *x*-*y* planes at *z/H* = 0.983.

Equation) described earlier in the works Launder and Spalding[9] and Spalding and Patankar[10] The turbulence characteristics were represented by a modified *k*-ε model to account for normal and shear stresses and near-wall functions, as previously described. Fluid properties such as densities, viscosity, and thermal conductivity were obtained from references. Kameel[1] introduced the partial differential equation of energy to the 3DHVAC program and solved the thermal airflow characteristics inside large rooms and in operating theaters, and used the hyperbolic grid generation equation and reported the airflow and heat transfer characteristics in similar configurations. It was found that airflow patterns in rooms are more quite using longer supply grilles, and both micro and macro mixing levels are influential. Neilson[11] had similar findings. More maximum absolute velocities and higher turbulent characteristics are demonstrated in situations with smaller supply jets for cold flows.

Good agreements were shown in Figures 6.4 and 6.5 between measured and predicted velocity components at various locations in the vicinity of the supply grilles. The corresponding temperature contour comparisons shown in Figures 6.6 and 6.7 also indicated good agreement to within 0.5°C. It can be concluded that the modeling capabilities can adequately predict the local flow pattern and heat transfer in air-conditioned spaces. Their predictions demonstrated flow and isothermal lines that are similar in form but differ in details at different operating conditions.

To achieve and maintain good IAQ conditions, it is necessary to remove or dilute airborne contamination in the enclosed space. A ventilation air distribution pattern has a great effect on the IAQ in enclosed spaces, especially healthcare applications. The primary tasks of a ventilation system are to remove the contaminated air from the room and supply the occupied region of the

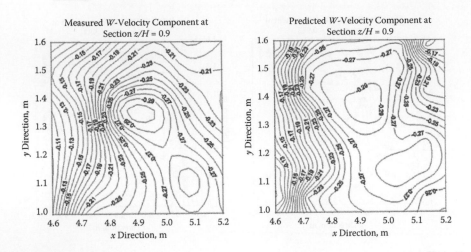

FIGURE 6.5
Comparisons between measured and predicted downward velocity components at horizontal *x-y* planes at *z/H* = 0.9.

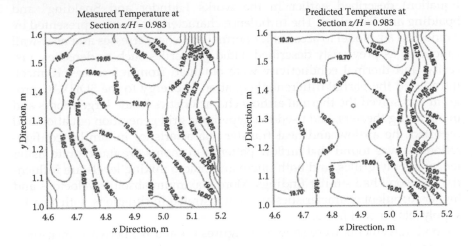

FIGURE 6.6
Comparisons between measured and predicted air temperatures at horizontal *x-y* planes at *z/H* = 0.983.

room with clean air. In confronting real-world computing problems, it is frequently advantageous to use several computing techniques synergistically rather than exclusively, resulting in the construction of complementary hybrid intelligent systems. Neuro-fuzzy systems are intelligent systems that combine knowledge, techniques, and methodologies from various sources. These intelligent systems are supposed to possess humanlike expertise within a specific domain, adapt themselves and learn to do better in changing environments, and explain how they make decisions or take actions. The quintessence of

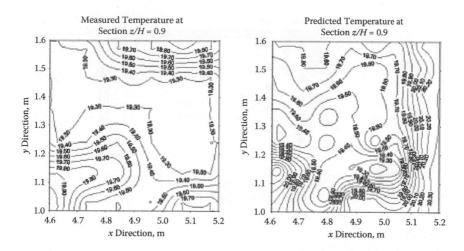

FIGURE 6.7
Comparisons between measured and predicted air dry-bulb temperature contours at horizontal *x-y* planes at $z/H = 0.9$.

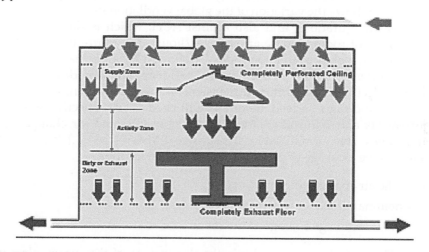

FIGURE 6.8
New hypothesis layout.

designing intelligent systems of this kind is neuro-fuzzy computing neural networks that recognize patterns and adapt themselves to cope with changing environments—fuzzy inference systems that incorporate human knowledge and perform inference and decision making. A new hypothesis was introduced by Kameel[1] (Figure 6.8) to evaluate the airflow regimes in real surgical operating theaters relative to the proposed ideal hypothesis.

An artificial neural network model (Figure 6.9) was introduced to be capable of evaluating and predicting the airflow movement and distribution efficiency. The present contribution should be useful to advise architectural

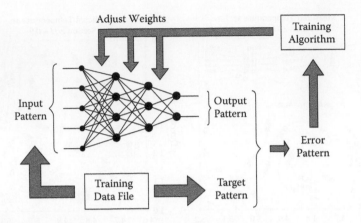

FIGURE 6.9
Artificial neural network diagram.

and HVAC design engineers (Figure 6.10) regarding the optimum approach to design for high ventilation efficiency.

Figure 6.10 shows the variation of the global ventilation index (GVI) versus the room configurations, and also against ACH, which is the air change rate (number of room volume air changes/hour). The figure indicates that increasing the global ventilation index is followed by a decrease in the air change efficiency, so the HVAC system designer will face difficulties in properly selecting the optimum dimensions of the supply plenum to attain the optimum design. Indeed, he or she will need to first specify the space requirements and reintroduce these in terms of GVI and air change efficiency values. The normalized local mean age (LMA) = $[\tau_p - \tau_{min}]/[\tau_{max} - \tau_{min}]$, where τ_p is the local age of particulates at a certain location.

E_{ac} is the air change efficiency = $\tau_n/2\tau_m$

τ_n = nominal time = enclosure volume/supply flow rate

τ_m = local mean age average of the room

From Figure 6.10a, one can obtain the dimensions of the supply plenum by the intersection between the GVI and the air change efficiency requirement. The region between the two values of GVI and air change efficiency will demarcate the possible dimensions that should be followed. The figure shows the effect of upper exhaust location height on the global ventilation effectiveness and air change efficiency. The difference between GVI1 and GVI2 is very small. GVI and air change efficiency increased simultaneously up to ($HUL/H = 0.667$), and then the air change efficiency decreases with the increase of (HUL/H). The range ($0.73 > HUL/H > 0.667$) is the most suitable position of the upper exhaust grille, for the present configuration and HVAC system design of Figure 6.10b. In general, one can observe that the most suitable place of the upper exhaust grille is close to the upper edge of the activity zone.

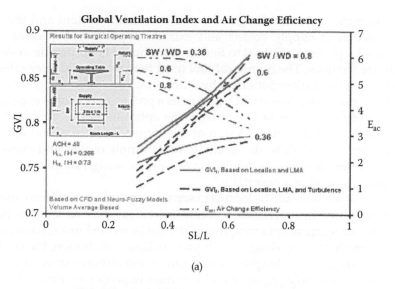

(a)

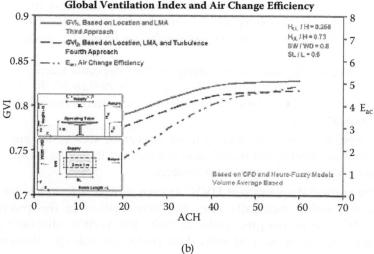

(b)

FIGURE 6.10
(a) Numerical and fuzziness results for global ventilation index versus room configurations.
(b) Numerical and fuzziness results for global ventilation index versus air change rates.

6.7 Conclusions

The air is not a medium only, but it is also a guard in critical applications. The airflow can be used as an engineering tool to provide a contaminant-free area. The proper direction of the airflow increases the possibilities of pollutant scavenging from healthcare applications. The proper airflow direction starts from the optimum design of the HVAC airside system and the optimum election of supply outlets and extraction ports. The present work introduced a preliminary trial to find the optimum HVAC airside design in healthcare facilities. The numerical tool used here was found very effective in predicting the airflow pattern in healthcare facilities with reasonable cost. So it is recommended to use CFD utilities as a preliminary tool to explore the optimum HVAC airside design. Indeed, the optimum HVAC airside design starts from the architectural design of healthcare facilities. Good architectural design allows HVAC designers to locate supply outlets and extraction ports in optimum locations. So the HVAC airside design should start from the planning phase of a hospital.

The critical areas should be located in separate sections or divisions. Indeed, in large hospitals, it is preferable to locate the critical areas on separate floors. Especially the surgical operating rooms should be located in a separate suite. For the already existing designs with poor airflow distribution, the environmental engineer in the hospitals can improve the airflow pattern by redistributing the medical equipment and furniture in proper locations.

From the numerical results obtained in the scope of this work that included airflow patterns, kinetic energy of turbulence, temperature, and relative humidity contours, the following summary of results can be used as design guides:

1. The IAQ in the surgical operating theaters depends mainly on the airflow distribution pattern. The optimum design of the airflow pattern can protect the surgery staff and the patients from the infection possibilities.

2. Indeed, the operating zone is most important area in the surgical operating theater, but that should not distract the focus away from other regions in the operating theaters.

3. The vertical displacement HVAC system design is found to be the most suitable, especially after the investigation of the practical HVAC system designs. Unfortunately, the vertical displacement design based on several individual perforated ceiling diffusers is also not recommended.

4. The extract system design should be developed on the basis of the two-level extract, one near the floor and the other at the end of the operating zone (i.e., about 2 m from the finished floor). Very few practice designs around the world insert the upper level of extract.

The air is not just a medium, but it can also be considered an effective tool in the hands of HVAC system designers of medical care facilities to provide a clean environment, and to provide the required protections for the surgery staff and patients. The fully vertical downward flow complete piston effect configuration can be considered the ideal design of the airflow distribution pattern to provide a clean operating theater. To reduce the infection possibilities of the surgery staff and patients, it should reduce the backward flow from the polluted zones toward the clean zones and recirculated zones. Air infection by direct and indirect ways should be avoided.

References

1. Kameel, R. 2002. Computer aided design of flow regimes in air-conditioned operating theatres. Ph.D. thesis, Cairo University, Egypt.
2. Kameel, R., and Khalil, E. E. 2003. Simulation of flow, heat transfer and relative humidity characteristics in air-conditioned surgical operating theatres. AIAA 2003-858. 41st Aerospace Sciences Meeting and Exhibit, Reno, NV.
3. Liu, Y., and Moser, A. 2002. Airborne particle concentration control for an operating room. ROOMVENT 2002, Denmark, p. 229.
4. Khalil, E. E. 2007. Numerical simulation of flow regimes and heat transfer interactions in complex geometries. AIAA-2007-4726.
5. Khalil, E. E. 2006. Flow regimes and thermal patterns in air conditioned operating theatres. *Proceedings Climamed 2006*, Lyon, November.
6. ASHRAE. 2009. *Fundamentals*. ASHRAE, Atlanta, GA.
7. Khalil, E. E., and Kameel, R. A. 2011. Experimental investigation of flow regimes in an operating theatre of 1200-beds teaching hospital. *Proceedings of ASHRAE APCBE 2011*, Jakarta, Indonesia, October.
8. Khalil, E. E. 2009. Thermal management in hospitals: Comfort, air quality and energy utilization. *Proceedings of ASHRAE, RAL*, Kuwait, October.
9. Launder, B. E., and Spalding, D. B. 1974. The numerical computation of turbulent flows. *Computer Methods in Applied Mechanics* 269–275.
10. Spalding, D. B., and Patankar, S. V. 1974. A calculation procedure for heat, mass and momentum transfer in three dimensional parabolic flows. *International Journal of Heat and Mass Transfer* 15, 1787.

Recommended Readings

Khalil, E. E. 2000. Computer aided design for comfort in healthy air conditioned spaces. In *Healthy Buildings 2000*, Finland, 2, 461–466.

Sandberg, M., and Sjoberg, M. 1983. The use of moments for assessing air quality in ventilated rooms. *Building and Environment* 18, 181–197.

Skaret, E., and Mathisen, H. M. 1983. Ventilation efficiency—A guide to efficient ventilation. *ASHRAE Transactions* 89, 480–495.

Peng, S., Holmberg, S., and Davidson, L. 1997. On the assessment of ventilation performance with the aid of numerical simulations. *Building and Environment* 32, 497–508.

Ross, A. D. 1999. On the effectiveness of ventilation. Ph.D. thesis, Eindhoven University of Technology, Eindhoven, Netherlands.

Sandberg, M. 1981. What is ventilation efficiency? *Building and Environment* 16, 123–135.

Deng, Q., and Tang, G. 2001. Ventilation effectiveness—Physical model and CFD solution. *Proceedings of the 4th International Conference on Indoor Air Quality, Ventilation and Energy Conservation in Buildings*, Changsha, Hunan, Hong Kong, China.

Han, X. V., and Chen, B. 2001. The interference of surrounding physical factors to IAQ subjective evolution. *Proceedings of the 4th International Conference on Indoor Air Quality, Ventilation and Energy Conservation in Buildings*, Changsha, Hunan, Hong Kong, China.

Zhu, C., Li, N., and Wen, W. 2001. Grey assessment of indoor air quality. *Proceedings of the 4th International Conference on Indoor Air Quality, Ventilation and Energy Conservation in Buildings*, Changsha, Hunan, Hong Kong, China.

Chow, T. T., Ward, S., Liu, J. P., and Chan, F. C. K. 2000. Airflow in hospital operating theatre: The Hong Kong experience. In *Proceedings of Healthy Buildings*, Finland, vol. 2.

Nelson, P. V. 1989. Numerical predictions of air distribution in rooms. *ASHRAE* 8, 31–38.

Kameel, R. 2002. Computer aided design of flow regimes in air-conditioned operating theatres. Ph.D. thesis, Cairo University.

Takagi, H., and Hayashi, I. 1991. NN-driven fuzzy reasoning. *International Journal of Approximate Reasoning*, 5(3), 191–212.

Sugeno, M., and Kang, G. T. 1988. Structure identification of fuzzy model. *Fuzzy Sets and Systems*, 28, 15–33.

Kondo, T. 1986. Revised GMDH algorithm estimating degree of the complete polynomial. *Transactions of the Society of Instruments and Control Engineers*, 22(9), 928–934.

Khalil, E. E. 2002. Energy efficiency in air-conditioned operating theatres. Paper 20050. IECEC 2002.

Khalil, E. E. 1978. Numerical procedures as a tool to engineering design. *Informatica* 78, 1–7.

Kameel, R., and Khalil, E. E. 2002. Predictions of flow, turbulence, heat transfer and humidity patterns in operating theatres. ROOMVENT 2002.

Sekhar, S. C., Wai, T. K., Cheong, D., and Hien, W. N. 2001. Indoor air quality, ventilation and energy studies in hot and humid climates. Clima 2000, Napoli.

Medhat, A. A. 1999. Air flow patterns in air conditioned rooms. Ph.D. thesis, Cairo University.

Khalil, E. E. 1999. Fluid flow regimes interactions in air conditioned spaces. In *Proceedings of the 3rd Jordanian Mechanical Engineering Conference*, Amman, May, p. 79.

Kameel, R., and Khalil, E. E. 2003. Simulation of flow, heat transfer and relative humidity characteristics in air-conditioned surgical operating theatres. AIAA-2003-0858. AIAA.

Khalil, E. E., Spalding, D. B., and Whitelaw, J. H. 1975. The calculation of two dimensional furnaces. *International Journal of Heat and Mass Transfer* 18, 775–791.

Spalding, D. B., and Patankar, S. V. 1974. A calculation procedure for heat, mass and momentum transfer in three dimensional parabolic flows. *International Journal of Heat and Mass Transfer* 15, 1787.

Launder, B. E., and Spalding D. B. 1974. The numerical computation of turbulent flows. *Computer Methods in Applied Mechanics* 3(2), 269–1289.

Khalil, E. E. 2010. Indoor air quality, airborne infection control and ventilation efficiency in hospital operating rooms. Topc-00075-2010. *Proceedings of ASHRAE_IAQ2010*, Malaysia, November.

Kameel, R., and Khalil, E. E. 2003. Air flow regimes in operating theatres for energy efficient performance. AIAA-2003-0686.

Huzayyin, O. A. S. 2005. Flow regimes and thermal patterns in air conditioned operating theatres. M.Sc. thesis, Cairo University.

Huzayyin, O. A. S., and Khalil, E. E. 2006. Applications of computational techniques to assess energy efficiency, air quality and comfort in air-conditioned spaces. AIAA-2006-1500.

Huzayyin, O. A. S., and Khalil, E. E. 2005. On the modelling of airflow regimes in surgical operating rooms. AIAA-2005-0753.

Nielsen, P. V. 1989. Numerical prediction of air distribution in rooms. ASHRAE, Building Systems: Room Air and Air Contaminant Distribution.

ASHRAE Standard 55-2010. 2010. ASHRAE, Atlanta, GA.

WHO. 1989. Indoor air quality: Organic pollutants. Report on a WHO meeting, Euro Report and Studies III, WHO Regional Office for Europe, Copenhagen.

Higgins, I. T. T. 1983. What is an adverse health effect? *APCA Journal* 33, 661–663.

Liviana, J. E., Rohles, F. H., and Bullock, P. E. 1988. Humidity comfort and contact lenses. *ASHRAE Transactions*, 94, 3–11.

Tanabe, S., Kimura, K., and Hara, T. 1987. Thermal comfort requirement during the summer season in Japan. *ASHRAE Transactions*, 93(1), 564–577.

Nevins, R., Gonzalez, R. R., Nishi, Y., and Gagge, A. P. 1975. Effect of changes in ambient temperature and level of humidity on comfort and thermal sensations. *ASHRAE Transactions*, 81(2), 150–157.

ASHRAE. 2009. *Fundamentals*. ASHRAE, Atlanta, GA.

ASHRAE Standard 62-2010. 2010. *Ventilation for acceptable indoor air quality*. ASHRAE, Atlanta, GA.

7

Examples of Typical Air Conditioning Projects

7.1 Environmental Control

When designing an efficient air conditioning system care should be exerted to satisfy the following requirements.

7.1.1 Temperatures

- The temperature should be controlled by change of supply temperature without any airflow control.
- Temperature difference between the warm and cool regions should be minimized to decrease the airflow drift.
- Good airflow distribution is required to create a homogenous domain without a large difference in the temperature distribution.
- Acceptable temperature in the occupancy zones as indicated in relevant codes.

7.1.2 Relative Humidity Control

Relative humidity affects human comfort directly and indirectly. It is a thermal sensation, skin moisture, discomfort, and tactile sensation of fabrics, health, and perception of air quality. Low humidity affects comfort and health. Comfort complaints about dry nose, throat, eyes, and skin occur in low-humidity conditions, typically when the dew point is less than 2°C. The upper-humidity limit was a dew point of 17°C in the ASHRAE standards, based not so much on comfort as on considerations of mold growth and other moisture-related phenomena.

At lower levels of humidity, thermal sensation is a good indicator of overall thermal comfort and acceptability. But at high humidity levels, thermal sensation alone is not a reliable predictor of thermal comfort. The most proper conditions are between 35 and 50% (Figure 7.1).

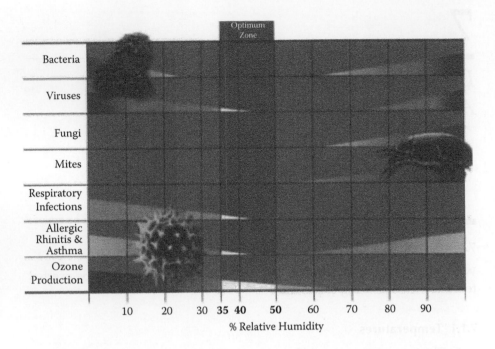

FIGURE 7.1
Relative humidity conditions.

7.1.3 Airflow Velocity Control

Airflow velocities should be kept to the minimum possible not to cause draft. However, the air velocities should also be high enough to enhance the heat transfer characteristics. For highly contaminated areas, the local velocity should be greater than or at least equal to 0.2 m/s, which has back influence on the value of the supplied air to overcome this condition. The unidirectional laminar airflow pattern is commonly attained at a velocity of 0.45 ± 0.10 m/s.

7.1.4 Ventilation

Ventilation rates shall be in accordance with ASHRAE Standard 62-2007.[1] Some facilities will have exhaust requirements that exceed required ventilation, and this will result in increasing outdoor air supply to balance the exhaust. Kitchen hood exhaust quantities dictate the quantity supply air required for air conditioning. To conserve energy 80% of the hood's air requirement is provided by a makeup air unit. A maximum of 50% recirculated air shall be transferred from the dining area to the kitchen and dishwashing areas. The dining areas shall be maintained at negative pressure relative to adjacent areas. The kitchen shall be maintained at negative pressure relative to the dining areas.

7.2 Air Filtration

An air filtration system will be designed to achieve indoor conditions comparable with ASHRAE standards for similar types of buildings. Because of the widely fluctuating levels of atmospheric particulate due to sandstorms and dust storms, a multistage filter system will be incorporated.

All air handling units will have three bases filtration stages:

1. The first stage will be an inertial separator installed on the outside air duct to remove sand dust particles from the outside air prior to entering the air handling units.
2. The second-stage filter located inside the air handling units will be of a pleated disposable type having an average efficiency of 30% based on ASHRAE Standard 52-2010.[2] The major advantages of the pleated filter are that replacement is required once annually compared to eight times for the flat filter under the same conditions.
3. The third-stage filter, also located in the air handling units, will be of a high-efficiency bag type, having an efficiency of 80–85% based on ASHRAE Standard 52-2010.[2]

This means:

1. The first and second stages will be sand trap louver installed on the outside air duct to remove sand.
2. In these two stages filters will hold any particulates up to an average efficiency of 30% based on ASHRAE Standard 52-2010.[2]
3. The third stage located inside the unit will be a bag type filter having an efficiency of 80–85% based on ASHRAE Standard 52-2010.[2]
4. Sand trap louvers shall be sized for a face velocity (200 to 300 fpm).
5. Additional filter stages may be added according to the zone application needs as described in a detailed manner in General and Design Bases of the Health Care Facilities.[1]

7.3 Pressure Relationships and Ventilation

Ventilation in accordance with ASHRAE Standard 62-2007,[1] *Ventilation for Acceptable Indoor Air Quality*, should be used for areas where specific

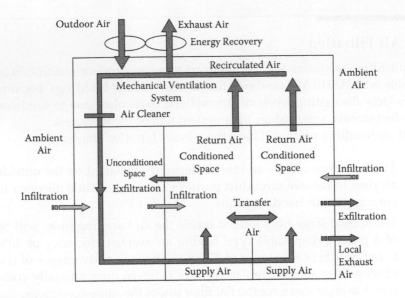

FIGURE 7.2
Pressure relationship.

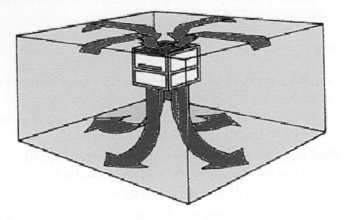

FIGURE 7.3
Airflow movement.

standards are not given. Where a higher outdoor air requirement is called for in ASHRAE Standard 62, the higher value should be used.

Design of a ventilation system must, as much as possible, provide air movement from clean to less clean areas. In critical care areas, constant volume systems should be employed to ensure proper pressure relationships (Figure 7.2) and ventilation, except in unoccupied rooms. Figure 7.3 shows the interactions between the various air-conditioned, ventilated spaces and the outdoors.

Air change per hour (ACH) plays an important role in providing a free contamination place and a reliable method to get rid of odors and bad smell.

The rooms are to usually be served by 2–6 ACH. Some critical rooms could be served by values up to 12 ACH. Actually, the proper value of the ACH should improve the airflow distribution in the space.

7.4 Air Movement

Undesirable airflow between rooms and floors is often difficult to control because of open doors, movement of staff and personnel, temperature differentials, and stack effect, which is accentuated by the vertical openings such as chutes, elevator shafts, stairwells, and mechanical shafts common to buildings. The effect of others may be minimized by terminating shaft openings in enclosed rooms and by designing and balancing air systems to create positive or negative air pressure within certain rooms and areas.

Systems serving highly contaminated areas should maintain a positive or negative air pressure within these rooms relative to adjoining rooms or the corridor. *The pressure is obtained by supplying less air to the area than is exhausted from it.* This induces a flow of air into the area around the perimeters of doors and prevents an outward airflow. A differential in air pressure shall be maintained only in an entirely closed room. Therefore, it is important to obtain a reasonably close fit of all doors or other barriers between pressurized areas. This is best accomplished by using weather stripping and drop bottoms on doors. The opening of a door or closure between two areas instantaneously reduces any existing pressure differential between them to such a degree that its effectiveness is nullified. When such openings occur, a natural interchange of air takes place because of the thermal currents resulting from temperature differences between the two areas. For areas requiring both the maintenance of pressure differentials to adjacent spaces and personnel movement between the critical area and adjacent spaces, the use of appropriate air locks or anterooms is indicated.

7.5 Air Quality

HVAC systems must provide air virtually free of dust, dirt, odor, and chemical and radioactive pollutants. In some cases, outside air may be hazardous to occupants with respiratory or pulmonary conditions. In such instances, systems that intermittently provide maximum allowable recirculated air should be considered. Air quality must also be maintained to provide a healthy, comfortable indoor environment. Sources of pollution exist in both the internal and external environments (Figure 7.4). The indoor air quality is

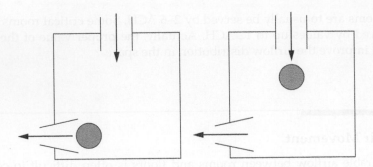

FIGURE 7.4
Pollutant tracking.

controlled by removal of the contaminant or by dilution. ASHRAE Standard 62-2007[1] prescribes both necessary quantities of ventilation for various types of occupancies and methods of determining the proportions of outside air and recirculated air. Although proper air conditioning designs are helpful in the prevention and treatment of diseases, the application of air conditioning to health facilities presents many specific problems. Those are not encountered in the conventional comfort conditioning design.[3] Mainly the contaminant distribution depends on the location of the pollutant source relative to the outlets in inlets of the airside system.

7.6 Smoke Control

As the ventilation design is developed, a proper smoke control strategy must be considered. Passive systems rely on fan shutdown, smoke and fire partitions, and operable windows. Proper treatment of duct penetrations must be observed. Active smoke control systems use the ventilation system to create areas of positive and negative pressures that, along with fire and smoke partitions, limit the spread of smoke. The ventilation system may be used in a smoke removal mode in which the products of combustion are exhausted by mechanical means.

7.7 Fire Control

Fire dampers will be installed in all supply, return, and exhaust air ducts where these ducts pass through 2 h rated firewalls. Smoke dampers will be installed in each duct penetration of 1 h rated smoke partitions and shafts. All fire and smoke dampers will be UL and NFPA approved.

TABLE 7.1

Noise Criteria in Spaces

Type of Area	RC or NC Criteria Range
Dormitories and apartments	30–35
Individual rooms or suites	30–35
Meeting/banquet rooms	30–35
Halls, corridors, lobbies, service/support areas	40–45
Executive offices and clinics	25–30
Conference rooms	25–30
Private area	30–35
Open offices and secretaries	35–40
Administrative assistant	35–40
Lobbies and circulation	35–40
Operating theaters	30–35
Computer/business machines	40–45
Public circulation	40–45
Bedrooms	30–35
Dining rooms	35–40
Athletic activity rooms	40–45

Note: The above values are for unoccupied spaces with all air conditioning systems operating.

7.8 Noise Criteria

The design will incorporate provisions for sound control in order to achieve an appropriate sound level for all activities and people involved. Generally, design goals for air conditioning system sound control for indoor areas will be in accordance with the ASHRAE *System* handbook,[4] as summarized in Table 7.1.

7.9 Factors Affecting System Selection

7.9.1 General Design Factors

In theory, if properly applied, every system can be successful in any building. However, in practice, such factors as initial and operating costs, space allocation, architectural design, location, and the engineer's evaluation and experience limit the proper choices for a given building type.

Heating and air conditioning systems should be

- Simple in design
- Of proper size for a given building
- Of generally fairly low maintenance
- Of low operating costs
- Of optimum inherent thermal control, as is economically possible

Such control might include materials with high thermal properties, insulation, and multiple or special glazing and shading devices. The relationship between the shape, orientation, and air conditioning capacity of a building should also be considered. Since the exterior load may vary from 30 to 60% of the total air conditioning load when the fenestration area ranges from 25 to 75% of the exterior envelope surface area, it may be desirable to minimize the perimeter area. For example, a rectangular building with a four-to-one aspect ratio requires substantially more refrigeration than a square building with the same floor area. Proper design also considers controlling noise and minimizing pollution of the atmosphere and water into which the system will discharge. The quality of the indoor air is also a major factor to consider in design.

Rehabilitation and retrofitting of existing buildings are also important parts of the construction industry because of increased construction costs and the necessity of energy consumption reduction. The HVAC system is often chosen by the owner and may not be based on an engineering study. To a great extent, system selection should be based on the engineer's ability to correlate those factors involving higher first cost or lower life cycle cost and benefits that have no calculable monetary value.

7.9.2 Load Characteristics

It is important to understand the load characteristics for the building to ensure that systems can respond adequately at part load as well as at full load. Systems must be capable of responding to large fluctuations in occupancy or to different occupancy schedules in different parts of the building. Areas in the building such as data or communications centers may have large 24 h equipment loads. Any building analyzed for heat recovery or total energy systems requires sufficient load profile and load duration information on all forms of building input to

1. Properly evaluate the instantaneous effect of one on the other when no energy storage is contemplated
2. Evaluate short-term effects (up to 48 h) when energy storage is used

Load profile curves consist of appropriate energy loads plotted against the time of day. Load duration curves indicate the accumulated number of hours at each load condition, from the highest to the lowest load for a day, a month, or a year. The area under load profile and load duration curves for corresponding periods is equivalent to the load multiplied by the time. These calculations must consider the type of air and water distribution systems in the building. Load profiles for two or more energy forms during the same operating period may be compared to determine load-matching characteristics under diverse operating conditions. For example, when thermal energy is recovered from a diesel electric generator at a rate equal to or less than the thermal energy demand, the energy can be used instantaneously, avoiding waste. But it may be worthwhile to store thermal energy when it is generated at a greater rate than demanded. A load profile study helps determine the economics of thermal storage.

Similarly, with internal source heat recovery systems, load matching must be integrated over the operating season with the aid of load duration curves for overall feasibility studies. Aside from environmental considerations, the economic feasibility of district heating and cooling systems is influenced by load density and diversity factors for branch feeds to buildings along distribution mains. For example, the load density or energy per unit length of distribution main can be small enough in a complex of low-rise, lightly loaded buildings located at a considerable distance from one another, to make a central heating, cooling, or heating and cooling plant uneconomical.

7.9.3 Design Factors

In many applications, design criteria are fairly evident, but in all cases, the engineer should understand the owner's and user's intent, because any single factor may influence system selection. The engineer's personal experience and judgment in the projection of future needs may be a better criterion for system design than any other single factor.

7.9.3.1 Comfort Level

Comfort, as measured by temperature, humidity, air motion, air quality, noise, and vibration, is not identical for all buildings, occupant activities, or use of space. The control of static electricity may be a consideration in humidity control. For spaces with a high population density, or with a sensible heat factor less than 0.75, a lower dry-bulb temperature reduces the generation of latent heat. Reduced latent heat may further reduce the need for reheat and save energy. Therefore, an optimum temperature should be determined.

7.9.3.2 Costs

Owning and operating costs can affect system selection and seriously conflict with other criteria. Therefore, the engineer must help the owner resolve these conflicts by considering factors such as the cost and availability of different fuels, the ease of equipment access, and maintenance requirements.

7.9.3.3 Local Conditions

Local, state, and national codes and regulations and environmental concerns must be considered in the design.

7.9.3.4 Automatic Temperature Control

Proper automatic temperature control maintains occupant comfort during varying internal and external loads. Improper temperature control may mean a loss of customers in restaurants and other public buildings. An energy management control system can be combined with a building automation system to allow the owner to manage energy, lighting, security, fire protection, and other similar systems from one central control system.

7.9.3.5 Fire, Smoke, and Odor Control

Fire and smoke can easily spread through elevator shafts, stairwells, ducts, and other means. An air conditioning system can spread fire and smoke by

1. Fan operation.
2. Penetrations required in walls or floors.
3. The stack effect without fan circulation; a properly designed and installed system can be a positive means of fire and smoke control.

An example of the schedule for office buildings is shown in Figure 7.5.

7.10 HVAC Life Cycle Cost Analyses

7.10.1 Introduction

The total cost is the combination of the owning, operating, and maintenance costs (Figure 7.6). The owning cost consists of the equipment and installation cost. The operating cost considered here would consist of the energy requirement costs, fuel, and water consumption of HVAC systems. The energy requirements also have an indirect impact on the environment. On the other hand, the maintenance cost estimation is separated from the operating cost analysis and includes the lifetime expectancy, spare parts cost estimation, and sudden breakdown repair cost estimation.

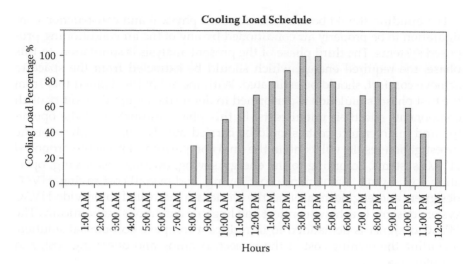

FIGURE 7.5
Office building load characteristics.

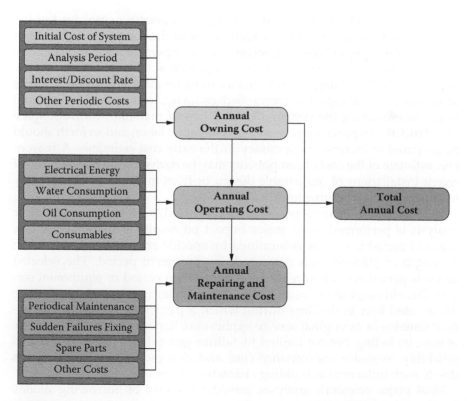

FIGURE 7.6
General procedure of analysis.

The building should be selected to have a physical and construction configuration to be properly air conditioned by any of the air conditioning proposed systems. The third phase of the present analysis is simulation. In this phase, the required energy, which should be extracted from the space to achieve comfort, should be estimated. With the aid of the defined tool from the first phase, simulation is performed to define the required quantity of all components and their initial costs. In this phase, estimation of the operating and maintenance costs should be carried out. The final analysis of the investigation results will contain the final comparisons among the proposed HVAC systems, according to the energy saving, owning cost saving, operating cost saving, maintenance cost saving, and overall cost saving. HVAC designers and engineers around the world are motivated to provide HVAC system designs to achieve comfort and hygiene conditions economically. The HVAC designer is demanded to introduce a complete economical solution, including the owning cost of the project, running and operating cost, and maintenance cost.

7.10.2 Owning Costs Analysis

Major decisions affecting annual owning and operating costs for the life of the building must generally be made prior to the completion of contract drawings and specifications. To achieve the best performance and economics, comparisons between alternative approaches to handle engineering requirements pertaining to each project must be made in the early stages of architectural design. Oversimplified estimates can lead to substantial errors in evaluating the system. Detailed lists of materials, controls, space and structural requirements, services, installation labor, and so forth should be prepared to increase the accuracy in the early cost estimates. A reasonable estimate of the cost of components may be derived from cost records of recent installations of comparable design or from quotations submitted by manufacturers and contractors.

The time frame (analysis period or study period) over which an economic analysis is performed has a major impact on results of the analysis. The analysis period is usually determined by specific analysis objectives, such as length of planned ownership or loan repayment period. The selected analysis period is often unrelated to depreciation period or equipment service life, although these factors may be important in the analysis. Service life as used here is the time during which a particular system or component remains in its original service application. Replacement may be for any reason, including, but not limited to, failure, general obsolescence, reduced reliability, excessive maintenance cost, and changed system requirements due to such influences as building characteristics or energy prices.

Most major economic analyses consider the cost of borrowing money (or the value of alternative uses of money), inflation, and the time value of money. Opportunity cost of money reflects the earnings that investing (or loaning) the money can produce. Time value of money reflects the fact that money received today is more useful than the same amount received a year from now, even with zero inflation, because the money is available for reinvestment earlier. Inflation (price escalation) decreases the purchasing or investing power (value) of future money because it can buy less (in the future). The cost or value of money must also be considered. When borrowing money, a percentage fee or interest rate must normally be paid. However, the interest rate may not necessarily be the correct cost of money to use in an economic analysis.

7.10.3 Operating Costs Analysis

Operating costs result from the actual operation of the system. They include fuel and electrical costs, wages, supplies, water, material, and maintenance parts and services. The ASHRAE handbook *Fundamentals*[3] outlines how fuel and electrical requirements are estimated. Note that total energy consumption cannot generally be multiplied by a per unit energy cost to arrive at annual utility cost. The total cost of electrical energy is usually a combination of several components: energy consumption charges, fuel adjustment charges, special allowances or other adjustments, and demand charges.

The power factor is the ratio of active (real) kilowatt power to apparent (reactive) kVA power. When calculating power bills, utilities should be asked to provide detailed cost estimates for various consumption levels. The final calculation should include any applicable special rates, allowances, taxes, and fuel adjustment charges. Electric rates may also have demand charges based on the customer's peak kilowatt demand. While consumption charges typically cover the operating costs of the utility, demand charges typically cover the owning costs. Demand charges may be formulated in a variety of ways: straight charge-cost per kilowatt per month, charged for the peak demand of the month; excess charge-cost per kilowatt above a base demand (e.g., 50 kW), which may be established each month; maximum demand (ratchet)-cost per kilowatt for the maximum annual demand, which may be reset only once a year (this established demand might either benefit or penalize the owner); and combination demand-cost per hour of operation of the demand. In addition to a basic demand charge, utilities may include further demand charges as demand-related consumption charges. The actual level of demand represents the peak energy use averaged over a specific period, usually 15, 30, or 60 min, accordingly; high electrical loads of only a few minutes' duration may never be recorded at the full instantaneous value. Alternatively, peak demand is recorded as the average of several consecutive short periods (i.e., 5 min out of each hour). The particular method of demand metering and billing is important when load shedding or shifting devices are considered. The portion of the total bill attributed to demand may vary widely, from 0% to as high as 70%.

7.10.4 Maintenance Costs Analysis

The quality of maintenance and maintenance supervision can be a major factor in the energy cost of a building. The ASHRAE handbook *Applications*[4] covers the maintenance, maintainability, and reliability of systems. The maintenance costs, based on data collected from U.S. markets, in 1983 U.S. dollars, showed a mean HVAC system maintenance cost of 32¢/ft² per year, with a median cost of 24¢/ft² per year. The age of the building has a statistically significant but minor effect on HVAC maintenance costs. When analyzed by geographic location, the data revealed that location does not significantly affect maintenance costs. Analysis also indicated that building size is not statistically significant in explaining cost variation.

The following method can be used for estimating and comparing the total building HVAC maintenance costs for various equipment combinations. (Manufacturers and other industry sources should be consulted for current information on new types of equipment.) This method assumes that the basic HVAC system in the building comprises fire tube boilers for heating equipment, centrifugal chillers for cooling equipment, and variable air volume (VAV) distribution systems. The total annual building HVAC maintenance cost for this system is 33.38¢/ft². Adjustment factors are then applied to this base cost to account for building age and variations in type of HVAC equipment as follows:

C, ¢/ft₂ = Total annual building HVAC maintenance cost (¢/ft₂) in 1983 dollars

= Base system maintenance costs + Age adjustment factor * Age in years
+ Heating system adjustment factor h + Cooling system adjustment factor c
+ Distribution system adjustment factor d

$$= 33.38 + 0.18n + h + c + d$$

Review Tables 7.2 and 7.3.

An example of cost analyses for an 800 tons refrigeration (TR) commercial building, five floors, total 2500 m², in Cairo is shown in Figure 7.7. Examples of the more efficient options are listed here and compared in Figure 7.8.

7.10.5 Design Specifications

Air systems

Constant air volume (CAV) system

Zoning

Mechanical cooling

Insulation

Outdoor intake

TABLE 7.2

Annual HVAC Maintenance Cost Adjustment Factors[a]

Age Adjustment	0.18n
Heating equipment h	
Water tube boiler	+0.77
Cast iron boiler	+0.94
Electric boiler	−2.67
Heat pump	−9.69
Electric resistance	−13.3
Cooling equipment c	
Reciprocating chiller a	−4.0
Absorption chiller (single stage)	+19.25
Water source heat pump	−4.72
Distribution system d	
Single zone	+8.29
Multizone	−4.66
Dual duct	−0.29
Constant volume	+8.81
Two-pipe fan coil	−2.77
Four-pipe fan coil	+5.80
Induction	+6.82

[a] In cents per square foot, 1983 U.S. dollars.

Note: These results pertain to buildings with older, single-stage absorption chillers. The data from the survey are not sufficient to draw inferences about the costs of HVAC maintenance in buildings equipped with new absorption chillers.

Exhaust outlets

Air filters

Vibration isolation

Ductwork systems

Ductwork materials

Acoustic treatment

7.11 Further Examples

Analyses of energy performance would be initiated from the power generation pattern in a country. In commercial buildings air conditioning systems can consume as much as 56% of the total energy used in the building. Therefore, it is a challenge to design an optimum HVAC airside system that

TABLE 7.3

Estimates of Service Lives of Various System Components

Equipment Category	Equipment Item	Median Years
Air conditioners	Window unit	10
	Residential single or split package	15
	Commercial through the wall	15
	Water-cooled package	15
Heat pumps	Residential air to air	15
	Commercial air to air	15
	Commercial water to air	19
Rooftop air conditioners	Single zone	15
	Multizone	15
Boilers, hot water (steam)	Steel water tube	24 (30)
	Steel fire tube	25 (25)
	Cast iron	35 (30)
	Electric	15
	Burners	21
Furnaces	Gas or oil fired	18
Unit heaters	Gas or electric	13
	Hot water or steam	20
Radiant heaters	Electric	10
	Hot water or steam	25
Air terminals	Diffusers, grilles, and registers	27
	Induction and fan coil units	20
	VAV and double-duct boxes	20
	Air washers	17
	Ductwork	30
	Dampers	20
Fans	Centrifugal	25
	Axial	20
	Propeller	15
	Ventilating roof mounted	20
Coils	Direct expansion (DX), water, or steam	20
	Electric	15
Heat exchangers	Shell and tube	24
	Reciprocating compressors	20
Package chillers	Reciprocating	25
	Centrifugal	25
	Absorption	25
Cooling towers	Galvanized metal	25
	Wood	20
	Ceramic	34
	Air-cooled condensers	20
	Evaporative condensers	20

TABLE 7.3 (*Continued*)

Estimates of Service Lives of Various System Components

Equipment Category	Equipment Item	Median Years
Insulation	Molded	20
	Blanket	24
Pumps	Base mounted	20
	Pipe mounted	10
	Sump and well	10
	Condensate	15
	Reciprocating engines	20
	Steam turbines	30
	Electric motors	18
	Motor starters	17
	Electric transformers	30
Controls	Pneumatic	20
	Electric	16
	Electronic	15
Valve actuators	Hydraulic	15
	Pneumatic	20

TABLE 7.4

Various Design Options

Option	Description
1	System with air-cooled chillers 2 × 400 TR
2	System with air-cooled chillers 4 × 200 TR
3	System with water-cooled chillers 2 × 400 TR
4	System with water-cooled chillers 4 × 200 TR
5	System with packaged water-cooled DX units and cooling towers

provides comfort and air quality in the air-conditioned spaces with efficient energy consumption. The conditions of the air to be maintained are dictated by the need for which the conditioned space is intended and comfort of users. So, the air conditioning embraces more than cooling or heating. Until now, the guidelines and design standards have not provide restricted utilization strategies of the conditioned air in the spaces. Indeed, this situation creates several inefficient systems, and consequently expensive energy invoices. In some critical facilities, such as hospitals, HVAC designers face the problem of balancing between attaining comfort conditions and efficient energy utilization. Table 7.4 demonstrates five different design options for a commercial building. The relationship between the HVAC system designs and the optimum conditions and optimum energy utilization is still under investigation today. In recent research, the effect of ventilation design on

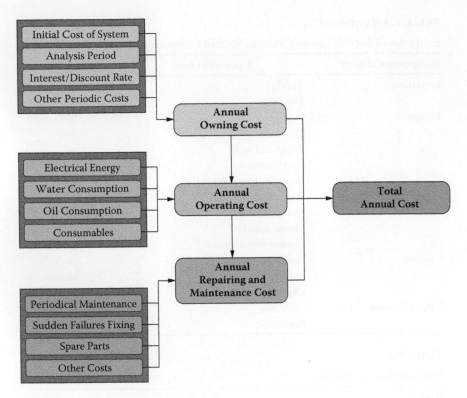

FIGURE 7.7
Total and initial cost analysis of 800 TR commercial building.

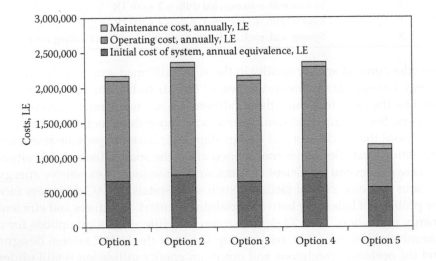

FIGURE 7.8
Cost comparisons of various options.

comfort and energy utilization has been investigated. A strong dependence relationship between the correct supplying conditions and comfort was found. Indeed, the displacement ventilation is recommended as an energy-efficient system. In recent years, new design traditions of the ventilation systems, such as under-floor ventilation systems, are growing to overcome the problems of the current systems. The under-floor air supply is recommended as an alternative to the ceiling air supply in office buildings to overcome the lack of flexibility in the ceiling systems and improve the comfort conditions. Actually, the energy utilization mainly depends on the optimum utilization of the conditioned air in the conditioned spaces.

The first example is shown here for a multipurpose hall.[5] The room dimensions are 12 m in length (L), 5.4 m in width (w), and 3.05 m in height (H). The forced air supply of cooled air streams out of high wall-mounted, 15° downward-inclined jets is investigated with mechanically extracted air from the top of the split units, as shown in Figure 7.9, with y being the vertical direction. This configuration played an important role in the main flow pattern and the creation of main recirculation zones. The internal obstacles would naturally obstruct the airflow pattern in different ways and means, by, for example, increasing the recirculation zones' size, relocating these, or deflecting the main airflow pattern. The local values of y^+ are about 12. Figure 7.10 illustrates the predicted vector plot at $x = 1.5$ m in a y-z plane. Strong recirculation zones were created and located in the vicinity of the furniture.

Experimental verifications were also reported by the author research group in a room provided by three commercially available high wall split units, each of 7.05 kW cooling at a supply airflow rate of 0.3 m³/s. The return air was taken back through a top horizontal return grille whose dimensions were 0.10 × 0.9 m. The supply-swiveled diffusers were at a fixed position during measurements. The present experiments were performed with the presence of the room furniture, such as the desk, tables, cupboard, and all additional

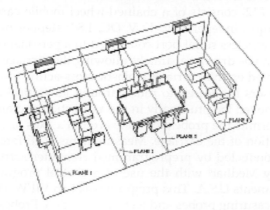

FIGURE 7.9
Multipurpose room configurations.

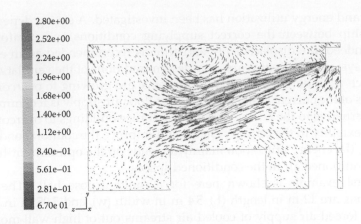

FIGURE 7.10
Predicted velocity vectors at a vertical plane, m/s, $x = 1.5$ m.

accessories, as shown in Figure 7.11a and b. Measurements were obtained at two vertical planes at the middle of the first two split air conditioners and also at a horizontal plane at the supply level of 2.85 m above the room finished floor. The local values of velocities, temperatures, and humidities were individually measured inside the room. The exterior walls were made of 0.4 m thick concrete brick wall and equipped with three high wall cooling systems.

A heated thermocouple anemometer was used. Its probe has a semisphere shape with a diameter of 1.0 mm, and it was supported in separate telescopic support of a cylindrical shape with a diameter equal to 12.0 mm at its base position. A test rig concept was designed and manufactured by Medhat.[6] The test rig was developed several times to improve and refine its accuracy. This system is capable of measuring in different spaces with different flooring types. The specific size limit of the effective used part of the test rig when cubed should never exceed 1:2000 of the volume of the room under test. The test rig (Figure 7.12) consists of a chained-wheel mobile carriage powered by two high-torque, independent, 12 V, DC, 1.80° stepper motors remotely controlled by a wireless serial port connected to a personal computer. Each stepper motor can be driven separately, allowing the carriage to move in the x-y plane. Mounted on this carriage, a plastic cross-acting tower carrying all probes and sensors is allowed to rotate around its vertical axes. Mounted probes can be moved longitudinally in the z direction and also rotate along its axes. This permits the probes to rotate around x axes or y axes, depending on the position of the carriage. Measurements at different locations in the room are conducted by preprogrammed computer software, designed and adopted by Medhat[6] with the use of a control program licensed by National Instruments U.S.A. This program was Lab VIEW. Figure 7.2A and B depicts the measuring probes and test rig in motion. Probes, when located at the desired positions, would measure data that are transferred instantaneously in batches to the computer for analysis; this is followed by a set of

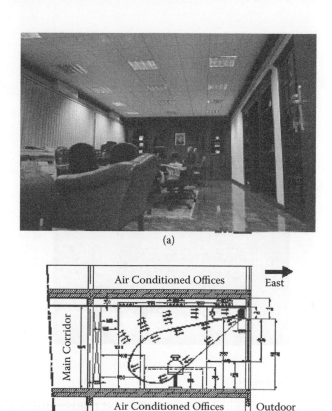

FIGURE 7.11
(a) Test room configuration. (b) Vertical cross-sectional view of the room.

measured data to the test rig again in a handshaking connection protocol. Data collected are instantaneously fed to the computer, recorded, and saved. Measurements of local velocities, humidity, and temperatures at any point in three-dimensional coordinates are based on the controlled volume of dimensions (0.4 m length, 0.5 m width, and 0.1 m height). During the room test procedure parts of the measured control volumes were overlapped by 50% of their volume to ensure accurate values (if required at critical expected zones). The distribution of the local velocity has been obtained with the aid of heated-thermocouple anemometer air velocity sensors. The distribution of the local mean temperature has been obtained with the aid of a thermal positive coefficient thermistor sensor and resistant temperature device (RTD). The local mean humidity distribution has been obtained by capacitive sensing elements. For the purpose of numerical simulation, the room was represented as indicated in Figure 7.9, with three wall-mounted air conditioners and with furniture fully modeled.

FIGURE 7.12
Test rig layout during motion.

Measurements of local velocities, humidity, and temperatures at any point in three-dimensional coordinates are based on controlled volume of dimensions (0.4 m length, 0.5 m width, and 0.1 m height). During the room test procedure, parts of the measured control volumes were overlapped by 50% of their volume to ensure accurate values (if required at critical expected zones). The distribution of the local velocity distribution has been obtained with the aid of heated-thermocouple anemometer air velocity sensors. The distribution of the local mean temperature has been obtained with the aid of a thermal positive coefficient thermistor sensor and RTD. The local mean humidity distribution has been obtained by capacitive sensing elements. Figure 7.13 presents a comparison between measured and predicted mean velocity profiles at $x = 4.2$ m in the room.

The scenarios of the numerical solution using commercial software included different cases related by number of occupants, location of A/C unit, locations of inlet (diffuser) and outlet (return), and the lighting. Practically, the poor zones shown in the next schemes do not exist clearly as introduced, because the diffuser of the air conditioner is distributing air from 0 to 90° and swimming all zones. From the previous results, one can conclude that the airside designs have a strong influence on the velocity and temperature distribution, and consequently on the indoor air quality (IAQ). The location of the supply outlets plays a major role in this distribution. The extraction ports should be located in the right location to correct the errors of the

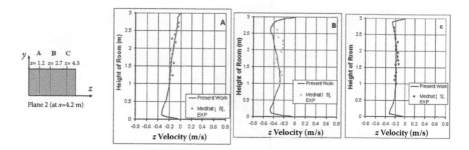

FIGURE 7.13

Comparisons between measured and predicted mean velocity profiles at x = 4.2 m. (From Aly, N. A., Improvement of Velocity and Temperature Fields in Air Conditioned Rooms Using CFD Modeling, MSc thesis, Cairo, Egypt, 2007.)

bad selection of the supply outlet positions. Due to the architectural design restrictions, designers may be forced to incorporate a certain design that is seen to yield better airflow, temperature, and relative humidity behavior.

From the previous results, one can conclude that the airside designs have a strong influence on the velocity and temperature distribution, and consequently on the emerging indoor air quality. The location of the supply outlets plays a major role in the distributions of flow and thermal energy. Air acts not only as an energy carrier but also as a means of protection of occupation zones from excessive drafts and accumulation of odors and void air currents. The extraction ports should be located in the right location to rectify the errors resulting from the bad and nonoptimum selection of the supply outlet positions. Such case is usually enforced by room interior decor or structural restrictions.

The second example[7] relates to the air-conditioned power transformer rooms where the temperature of transformer surfaces should be kept low for better performance. The main switchgear room is part of the transformer station complex having principal dimensions of 17 m long × 14 m wide × 6 m high. It includes mainly four dry type power transformers, as shown in Figure 7.14. The room air conditioning system comprises four supply openings in the circular supply duct and an equal number of return/exhaust grilles. The main heat source in the room is the transformers in operation. The transformers were identical units that have the dimensions of 1.6 m long × 2 m wide × 2 m high.

Figure 7.15 denotes the predicted thermal contours where the observed asymmetries can be well attributed to the detailed structure of the transformer housing, which is still more complicated to be adequately represented by conventional numerical simulation.

The third case is for a movie theater[8,9] whose dimensions are 30.0 m long, 20.0 m wide, with varied ceiling height of 10.0 m at the front to 8.0 m at the rear; here Z is the vertical direction. Modeling of the effect of audience presence inside this theater was developed with full audience loads, which are

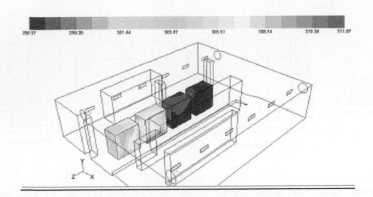

FIGURE 7.14
Predicted surface temperature of transformers; two of the transformers are out of service.

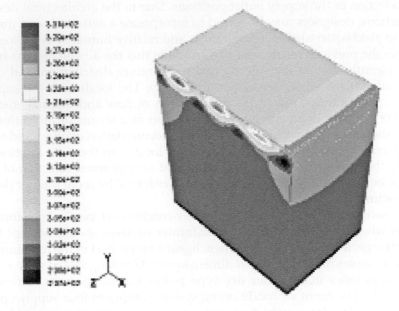

FIGURE 7.15
Predicted temperature contours, K, at the transformer casing.

indicated on the floor and balcony regions as shown in Figure 7.16. The present work describes the airflow characteristics for a given design that incorporated 47 ceiling-mounted air supply diffusers (29 of which are mounted at the main ceiling plus 18 diffusers mounted at the balcony floor level and 18 extract ports on both side walls).

Predicted flow patterns are shown in Figure 7.17 and demonstrate the level of details one might get with computational techniques. The merits and assets of computationally testing the various optional locations of ceiling supply grilles and their numbers favor computational techniques

9 Supply Ports

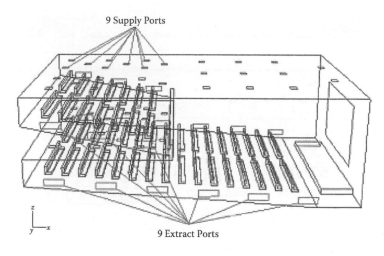

9 Extract Ports

FIGURE 7.16
The theater and seated audience modeling.

| 0 | 0.25 | 0.5 | 0.75 | 1 | 1.25 | 1.5 | 1.75 | 2 | 2.25 | 2.5 |

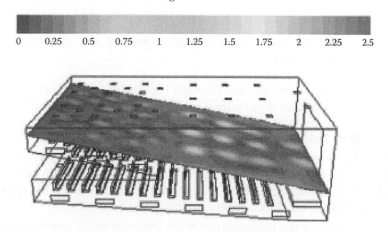

FIGURE 7.17
Velocity distribution at 1.7 from balcony floor.

as an affordable design tool for optimization of energy demands in large spaces, by simply better utilizing the energy embedded in the conditioned air. Naturally increasing the number of air return grilles would lead to more uniform airflow distribution and minimize the stagnant air zones. The corresponding CO_2 predictions are shown in Figure 7.18 at 1.8 m from the balcony floor. The results shown in the present work indicate a good agreement between the average predicted air velocity, average temperature, average relative humidity, and average CO_2 concentration with the ASHRAE recommended comfort[3] conditions within the occupied zones in the theater. The numerical algorithm utilized in the above three cases was run to yield grid-independent results with grids of the order of 1,000,000 nodes.[10]

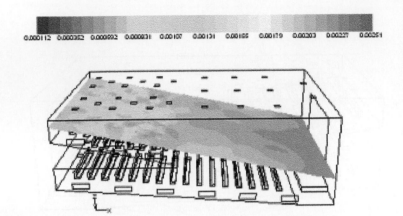

FIGURE 7.18
CO_2 distributions at 1.7 from balcony floor.

7.12 Engineering Tips for Energy-Efficient Buildings

The following tips represent the average more general and effective tips and hints to energy managers:

1. Define energy management through typical operation and maintenance perceptions; it is really a management task (data, communication, implementation, tracking); it is an art. Energy management should include operational efficiency and energy efficiency.

2. Determine available data, such as metered data, databases, reporting tools, utility bills, equipment energy efficiency/performance, direct digital control/energy management systems, available operation manuals, and building energy performance.

3. Select your energy management team who can easily communicate with management/supervisor staff. Bear in mind that the operations staff are critical; therefore, facilities and maintenance staff need to be involved. Develop lines of communication and evaluate available staff resources, capabilities, and required training,

4. Involve operators and maintenance team on the location. Solicit their ideas; they know their systems best and already often have good energy efficiency ideas. Determine what would be required to implement an electrical efficiency measure. Determine low- to no-cost options and evaluate ability to not sacrifice comfort or productivity while implementing an energy efficiency measure, and then identify how to make an energy efficiency measure institutionalized.

5. Select priority for energy saving through HVAC control, lighting control, process systems, chillers, and boilers. For that use, potential

team members may include energy, utilities, facilities, operation and maintenance, and end user stakeholders. Usually inefficiencies are not detectable by typical operation and maintenance practices. Needed is someone experienced with implementing energy efficiency procedures to facilitate the process. Gather information and data, turn off what you can, and if you can't turn it off, determine if you can reduce it. Set up a mechanism to facilitate communication, do not walk away, and set up a mechanism to continually apply energy efficiency measures.

6. It is imperative to involve operations and facility staff on the team and attempt to understand their perspective. Make it clear that suggested improvements are not a reflection of their current performance. Usually address concerns that changes will result in more "trouble calls," and show them the potential and actual results of the proposed changes improving energy efficiency indicators.

7. Measure or estimate how much energy is being used by different equipment, utilizing available meters and measurement instruments. A need emerges to determine if operational manuals are available or if operational knowledge has been institutionalized. The energy efficiency performance of facilities is to be evaluated through the knowledge of energy use data per gross floor area or other means to establish a rating/comparison; careful analyses of simulation data or engineering calculations should be followed.

8. Implement energy efficiency operational ideas such as the low- to no-cost ideas, maintain communication among people involved, and measure performance of the operational changes.

9. Measure where needed to determine savings from operational changes, in case you already have meters and sufficient measurement tools. Determine if the data are being properly evaluated. Continually evaluate energy data and energy efficiency operational changes and document any successes and savings.

10. Operational changes need to be continually implemented and properly documented and retained, especially an implementation plan to track key energy efficiency operational changes. Maintain communication between the energy manager, operators, maintenance, and management, and document savings to justify the energy efficiency program.

7.13 Conclusions

The reported comparisons between the present predictions of air velocities, temperatures, relative humidity, and carbon dioxide concentrations and the corresponding experimentations indicated qualitative agreements. The design configurations shown in this chapter yielded better air supply and return air utilization, which yielded the predicted CO_2 concentration within the occupied zones that are below the maximum allowable limit (1000 ppm). The three different cases were carefully selected to demonstrate the applicability and to indicate how to conserve energy simply by properly managing the air distributions. It is believed that for the purpose of the energy-efficient operation of air conditioning and other systems in large buildings, the energy manager should take into account the tips that are set out in this work. Different actors need different information. For giving relevant advice to the property owner regarding which measures are cost-effective, a very careful examination and calculation of the building's energy balance is necessary. A careful analysis is also necessary to give relevant information to the users on how they can decrease their energy use without decreasing, under an acceptable level, the indoor air quality and thermal comfort.

References

1. ASHRAE 62-2007. 2007. *Ventilation for acceptable indoor air quality*. ASHRAE, Atlanta, GA.
2. ASHRAE 55-2010. 2010. *Thermal comfort standard*. ASHRAE, Atlanta, GA.
3. ASHRAE. 2009. *Fundamental*. ASHRAE, Atlanta, GA.
4. ASHRAE. 2011. *Systems*. ASHRAE, Atlanta, GA.
5. Aly, N. A. 2007. Improvement of velocity and temperature fields in air conditioned room using CFD modelling. MSc thesis, Cairo University, Cairo, Egypt.
6. Medhat, A. A. 1999. Optimizing room comfort using experimental and numerical modeling. PhD thesis, Cairo University, Cairo, Egypt.
7. ElBialy, E. M., and Khalil, E. E. 2008. Experimental and computational investigations of air flow and thermal patterns in air conditioned power transformers room. AIAA-2008-5648. *Proceedings of the 6th IECEC*, Cleveland, OH, July.
8. Abdel-Samee, W. 2007. Numerical investigations of flow patterns and thermal comfort in air-conditioned cinema theatres. MSc thesis, Cairo University, Cairo, Egypt.
9. Khalil, E. E., and Abdel-Maksoud, W. 2009. Indoor air quality inside an air-conditioned large auditorium. Paper 348. *Proceedings of Healthy Buildings*, September.
10. Fluent. 2005. *CFD code manual*.

8

Indoor Environmental Quality

Indoor environmental quality, as the name implies, simply refers to the quality of the air in an office or other building environment. Workers are often concerned that they have symptoms or health conditions from exposure to contaminants in the building where they work. One reason for this concern is that their symptoms often get better when they are not in the building. While research has shown that some respiratory symptoms and illnesses can be associated with damp buildings, it is still unclear what measurements of indoor contaminants show that workers are at risk for disease. In most instances where a worker and his or her physician suspect that the building environment is causing a specific health condition, the information available from medical tests and tests of the environment is not sufficient to establish which contaminants are responsible. Indoor environments are highly complex, and building occupants may be exposed to a variety of contaminants (in the form of gases and particles) from office machines, cleaning products, construction activities, carpets and furnishings, perfumes, cigarette smoke, water-damaged building materials, microbial growth (fungal/mold and bacterial), insects, and outdoor pollutants. Other factors, such as indoor temperatures, relative humidity, and ventilation levels, can also affect how individuals respond to the indoor environment. Understanding the sources of indoor environmental contaminants and controlling them can often help prevent or resolve building-related worker symptoms. Practical guidance for improving and maintaining the indoor environment is available. Workers who have persistent or worsening symptoms should seek medical evaluation to establish a diagnosis and obtain recommendations for treatment of their condition.[1-7]

This section specifies requirements for indoor environmental quality, including indoor air quality, environmental tobacco smoke control, outdoor air delivery monitoring, thermal comfort, building entrances, acoustic control, daylighting, and low-emitting materials.

8.1 Definitions

The following words and terms shall, for the purposes of this chapter and as used elsewhere in this book, have the meanings shown herein:

Agrifiber products: Agrifiber products include wheat board, strawboard, panel substrates, and door cores. They do not include furniture, fixtures, and equipment (FF&E) not considered base building elements.

Composite wood products: Composite wood products include hardwood plywood, particleboard, and medium-density fiberboard. Composite wood products do not include hardboard, structural plywood, structural panels, structural composite lumber, oriented strand board, glued laminated timber as specified in *Structural Glued Laminated Timber* (ANSI A190.1-2002), or prefabricated wood I-joists.

Daylit area: Area under horizontal fenestration (skylight) or adjacent to vertical fenestration (window) as shown in Figures 8.1 and 8.2.

Daylit space: Space bounded by vertical planes rising from the boundaries of the daylit area on the floor to above the floor or roof.

HVAC units, small: Those containing less than 0.22 kg of refrigerant.

Interior, building: The inside of the weatherproofing system.

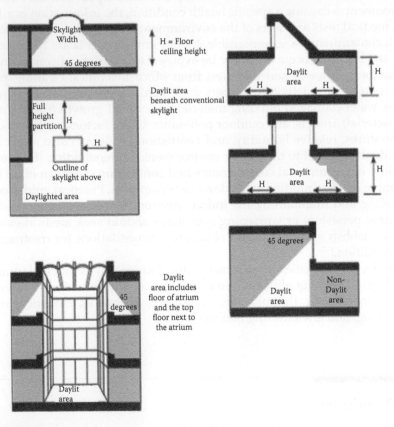

FIGURE 8.1
Horizontal daylit area.

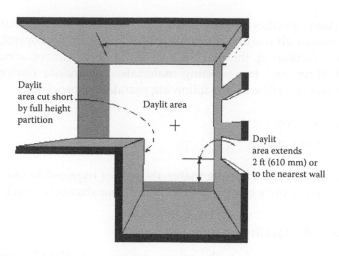

FIGURE 8.2
Vertical daylit area.

MERV: Filter minimum efficiency reporting value, based on ASHRAE 52.2-2009.

Moisture content: The weight of the water in wood expressed in percentage of the weight of the oven-dry wood.

Multioccupant spaces: Indoor spaces used for presentations and training, including classrooms and conference rooms.

Single-occupant spaces: Private offices, workstations in open offices, reception workstations, and ticket booths.

VOC: A volatile organic compound broadly defined as a chemical compound based on carbon chains or rings with vapor pressures greater than 0.1 mm of mercury at room temperature. These compounds typically contain hydrogen and may contain oxygen, nitrogen, and other elements.

8.2 IEQ Mandatory Provisions

Indoor environment quality is generally dictated by the following parameters:

1. Pollutant control
2. Indoor moisture control
3. Indoor air quality and exhaust
4. Environmental comfort
5. Visual comfort
6. Outdoor air quality

This section specifies requirements for indoor environmental quality, including indoor air quality, environmental tobacco smoke control, outdoor air delivery monitoring, thermal comfort, building entrances, acoustic control, daylighting, and low-emitting materials. The indoor environmental quality should comply with the following mandatory provisions:

1. Prescriptive option
2. Performance option

Daylighting and low-emitting materials are not required to use the same option, i.e., prescriptive or performance, for demonstrating compliance.

8.2.1 Indoor Air Quality

The building shall comply with various sections of ASHRAE Standard 62.1 with the following modifications and additions.

8.2.2 Minimum Ventilation Rates

1. The ventilation rate procedure of ASHRAE Standard 62.1 shall be used to design each mechanical ventilation system in the building.
2. The zone-level design outdoor airflow rates in all occupiable spaces shall be greater than or equal to the airflow calculated using the ventilation rate procedure in Section 6.2 of ASHRAE Standard 62.1.
3. The system-level design outdoor airflow rate calculation (Sections 6.2.3 to 6.2.5 of ASHRAE Standard 62.1) shall be based on the zone-level design outdoor airflow rates.

8.2.3 Outdoor Air Delivery Monitoring

8.2.3.1 Spaces Ventilated by Mechanical Systems

For variable air volume supply systems a permanently mounted, direct total outdoor airflow measurement device shall be provided that is capable of measuring the system minimum outdoor airflow rate. The device shall be capable of measuring flow within an accuracy of ±15% of the minimum outdoor airflow rate. The device shall also be capable of being used to alarm the building operator or for sending a signal to a building central monitoring system when flow rates are not in compliance.

Exceptions to Section 8.2.3.1 include the following:

1. For each air handling system that serves only densely occupied spaces, a direct total outdoor airflow measurement device is not required if a

permanently installed carbon dioxide (CO_2) monitoring system is provided for all densely occupied spaces. For air handling systems that serve only a single densely occupied space, CO_2 monitoring is required only for that space. The CO_2 monitoring system shall record ventilation system performance in terms of differential indoor-to-outdoor CO_2 levels. The CO_2 monitoring system shall be capable of indicating the CO_2 level in a direct readout display in the occupied space, conveying such level to a building central monitoring system, or both.

2. Constant volume air supply systems that use a damper position feedback system are not required to have a direct total outdoor airflow measurement device.

8.2.3.2 Naturally Ventilated Spaces

A permanently installed CO_2 monitoring system shall be provided in densely occupied spaces designed to operate without a mechanical ventilation system for any period of time that the space is occupied. Indoor CO_2 sensors or air sampling probes shall be located within the room between 1 and 2 m above the floor and on a wall location at least 6 m from operable openings. The CO_2 monitoring system shall be capable of indicating the CO_2 level in a direct readout display in the occupied space, conveying such level to a building central monitoring system, or both. The CO_2 sensors shall meet the requirements stated in Section 8.2.3.3. Where floor plans are less than 12 m wide, sensors shall be located as close to the center of the space as practical. One CO_2 sensor is allowed to be used to represent multiple spaces if the natural ventilation design uses passive stack(s) or other means to induce airflow through those spaces equally and simultaneously without intervention by building occupants.

8.2.3.3 CO_2 Sensors

Spaces with CO_2 sensors or air sampling probes leading to a central CO_2 monitoring station shall have one sensor or probe for each 1000 m^2 of floor space and shall be located in the room between 1 and 2 m above the floor. CO_2 sensors must be accurate to ± 50 ppm at 1000 ppm. For all spaces with CO_2 sensors, target concentrations shall be calculated for full and part load conditions (to include no less than 25, 50, 75, and 100%). Metabolic rate, occupancy, ceiling height, and outdoor air CO_2 concentration assumptions and target concentrations shall be shown in the design documents. Outdoor air CO_2 concentrations shall be determined by one of the following:

1. CO_2 concentration shall be assumed to be 400 ppm without any direct measurement.
2. CO_2 concentration shall be dynamically measured using a CO_2 sensor located near the position of the outdoor air intake.

8.2.4 Filtration and Air Cleaner Requirements

8.2.4.1 Particulate Matter

1. The particulate matter filters or air cleaners shall have a MERV of not less than 8 and shall comply with and be provided where required in Section 5.9 of ASHRAE Standard 62.1.

2. In addition to ASHRAE Standard 62.1, Section 6.2.1.1, when the building is located in an area that is designated "nonattainment" with the National Ambient Air Quality Standards for PM2.5 as determined by the AHJ (in the United States, by USEPA), particle filters or air-cleaning devices having a MERV of not less than 13 when rated in accordance with ASHRAE Standard 52.2 shall be provided to clean outdoor air prior to its introduction to occupied spaces.

8.2.4.2 Ozone

In addition to ASHRAE Standard 62.1, Section 6.2.1.2 requirements, when the building is located in an area that is designated "nonattainment" with the National Primary and Secondary Ambient Air Quality Standards for ozone as determined by the AHJ (in the United States, by USEPA), air cleaner cleaning devices having a removal efficiency of no less than the one efficiency specified in ASHRAE Standard 62.1, Section 6.2.1.2 shall be provided to clean outdoor air prior to its introduction to occupied spaces.

8.2.4.3 Bypass Pathways

All filter frames, air cleaner racks, access doors, and air cleaner cartridges shall be sealed to eliminate air bypass pathways.

8.2.5 Environmental Tobacco Smoke

1. Smoking shall not be allowed inside the building. Signage stating such shall be posted within 3 m of each building entrance.

2. Any exterior designated smoking areas shall be located a minimum of 7.5 m away from building entrances, outdoor air intakes, and operable windows.

3. Section 6.2.9 of ASHRAE Standard 62.1 shall not apply.

8.2.6 Building Entrances

All building entrances shall employ an entry mat system that shall have a scraper surface, an absorption surface, and a finishing surface. Each surface shall be a minimum of the width of the entry opening, and the minimum length is measured in the primary direction of travel.

Exceptions to Section 8.2.6 include the following:

1. Entrances to individual dwelling units.
2. Length of entry mat surfaces is allowed to be reduced due to a barrier, such as a counter, partition, or wall, or local regulations prohibiting the use of scraper surfaces outside the entry. In this case entry mat surfaces shall have a minimum length of 1 m of indoor surface, with a minimum combined length of 2 m.

8.2.6.1 Scraper Surface

The scraper surface shall comply with the following:

1. Shall be the first surface stepped on when entering the building
2. Shall be either immediately outside or inside the entry
3. Shall be a minimum of 1 m long
4. Shall be either permanently mounted grates or removable mats with knobby or squeegee-like projections

8.2.6.2 Absorption Surface

The absorption surface shall comply with the following:

1. Shall be the second surface stepped on when entering the building
2. Shall be a minimum of 3 ft (1 m) long, and made from materials that can perform both a scraping action and a moisture wicking action

8.2.6.3 Finishing Surface

The finishing surface shall comply with the following:

1. Shall be the third surface stepped on when entering the building
2. Shall be a minimum of 4 ft (1.2 m) long, and made from material that will both capture and hold any remaining particles or moisture

8.3 Thermal Environmental Conditions for Human Occupancy Comfort

The building shall be designed in compliance with ASHRAE Standard 55, Sections 6.1 (Design) and 6.2 (Documentation).

An exception to Section 8.3 is spaces with special requirements for processes, activities, or contents that require a thermal environment outside that which humans find thermally acceptable, such as food storage, natatoriums, shower rooms, saunas, and drying rooms.

8.4 Acoustical Control

8.4.1 Exterior Sound

Wall and roof-ceiling assemblies that are part of the building envelope shall have a composite outdoor-indoor transmission class (OITC) rating of 40 or greater or a composite sound transmission class (STC) rating of 50 or greater, and fenestration that is part of the building envelope shall have an OITC or STC rating of 30 or greater for any of the following conditions:

1. Buildings within 300 m of expressways
2. Buildings within 5 mi (8 km) of airports serving more than 10,000 commercial jets per year
3. Where yearly average day-night average sound levels at the property line exceed 65 dB

An exception to Section 8.4.1 is buildings that may have to adhere to functional and operational requirements such as factories, stadiums, storage, enclosed parking structures, and utility buildings.

8.4.2 Interior Sound

Interior wall and floor-ceiling assemblies separating interior rooms and spaces shall be designed in accordance with all of the following:

1. Wall and floor-ceiling assemblies separating adjacent dwelling units, dwelling units and public spaces, adjacent tenant spaces, tenant spaces and public places, and adjacent classrooms shall have a composite STC rating of 50 or greater.
2. Wall and floor-ceiling assemblies separating hotel rooms, motel rooms, and patient rooms in nursing homes and hospitals shall have a composite STC rating of 45 or greater.
3. Wall and floor-ceiling assemblies separating classrooms from restrooms and showers shall have a composite STC rating of 53 or greater.
4. Wall and floor-ceiling assemblies separating classrooms from music rooms, mechanical rooms, cafeteria, gymnasiums, and indoor swimming pools shall have a composite STC rating of 60 or greater.

8.4.3 Outdoor–Indoor Transmission Class and Sound Transmission Class

OITC values for assemblies and components shall be determined in accordance with ASTM E1332. STC values for assemblies and components shall be determined in accordance with ASTM E90 and E413.

8.5 Daylighting by Top Lighting

There shall be a minimum fenestration area providing daylighting by top lighting for large enclosed spaces. In buildings three stories and less above grade, conditioned or unconditioned enclosed spaces that are greater than 2000 m² directly under a roof with a finished ceiling height greater than 4 m and having a lighting power allowance for general lighting equal to or greater than 5.5 W/m² shall comply with the following.

8.5.1 Minimum Daylight Zone by Top Lighting

A minimum of 50% of the floor area directly under a roof in spaces with a lighting power density or lighting power allowance greater than 5 W/m² shall be in the daylight zone. Areas that are daylit shall have a ratio of minimum top lighting area to daylight zone area as shown in Table 8.1. For purposes of compliance with Table 8.1, the greater of the space lighting power density and the space lighting power allowance shall be used.

8.5.2 Skylight Characteristics

Skylights used to comply with Section 3.4.1 shall have a glazing material or diffuser that has a measured haze value greater than 90%, tested according to ASTM D1003 (notwithstanding its scope) or other test method approved by the authority having jurisdiction.

TABLE 8.1

Minimum Top Lighting Area

General Lighting Power Density (LPD) or Lighting Power Allowances in Daylight Zone	
W/ft² (W/m²)	Ratio of Minimum Top Lighting Area to Daylight Zone Area
14 W/m² < LPD	3.6%
10 W/m² < LPD < 14 W/m²	3.3%
5 W/m² < LPD < 10 W/m²	3.0%

Exceptions to Section 8.5.2 include the following:

1. Skylights with a measured haze value less than or equal to 90% whose combined area does not exceed 5% of the total skylight area.
2. Tubular daylighting devices having a diffuser.
3. Skylights that are capable of preventing direct sunlight from entering the occupied space below the well during occupied hours. This shall be accomplished using one or more of the following:

   - Orientation
   - Automated shading or diffusing devices
   - Diffusers
   - Fixed internal or external baffles
   - Airline terminals, convention centers, and shopping malls

8.6 Isolation of the Building from Pollutants in Soil

Building projects that include construction or expansion of a ground-level foundation and which are located on brownfield sites or in zone 1 counties identified to have a significant probability of radon concentrations higher than 4 pCi/L on the USEPA map of radon zones shall have a soil gas retarder system installed between the newly constructed space and the soil.

8.7 Prescriptive Option

8.7.1 Daylighting by Side Lighting

8.7.1.1 Minimum Effective Aperture

Office spaces and classrooms shall comply with the following criteria:

1. All north-, south-, and east-facing facades for those spaces shall have a minimum effective aperture for vertical fenestration (EAvf) as prescribed in Table 8.2.
2. Opaque interior surfaces in daylight zones shall have visible light reflectances greater than or equal to 80% for ceilings and 70% for partitions higher than 1.5 m in daylight zones.

TABLE 8.2

Minimum Effective Aperture for Side Lighting by Vertical Fenestration

Climate Zone	Minimum Effective Aperture for Side Lighting by Vertical Fenestration
1, 2, 3A, 3B	0.10
3C, 4, 5, 6, 7, 8	0.15

Exceptions to Section 8.7.1.1 include the following:

1. Spaces with programming that requires dark conditions (e.g., photographic processing).
2. Daylight zones where the height of existing adjacent structures above the window is twice the distance between the window and the adjacent structures.
3. Facades that are less than 10 ft (3 m) from an adjacent building. (For a space with multiple facades, those portions of other facades that do not qualify under this exception shall comply with Section 8.7.1.1.)

8.7.1.2 Office Space Shading

Each west-, south-, and east-facing facade shall be designed with a shading projection factor. The projection factor shall be no less than 0.5 that specified in Section 7.4.2.5. Shading is allowed to be external or internal using the projection factor, interior. The building is allowed to be rotated up to 45° for purposes of calculations and showing compliance. The following shading devices are allowed to be used:

1. Louvers, sun shades, light shelves, and any other permanent device. Any vertical fenestration that employs a combination of interior and external shading is allowed to be separated into multiple segments for compliance purposes. Each segment shall comply with the requirements for either external or interior projection factor.
2. Building self-shading through roof overhangs or recessed windows.
3. External buildings and other permanent infrastructure or geological formations that are not part of the building. Trees, shrubs, or any other organic shading device shall not be used to comply with the shading projection factor requirements.

Exceptions to Section 8.7.1.2 include the following:

1. Building projects that comply with the prescriptive compliance option. Translucent panels and glazing systems with a measured haze value greater than 90%, tested according to ASTM D1003

(notwithstanding its scope) or other test method approved by the authority having jurisdiction, and that are entirely 2.5 m above the floor, do not require external shading devices.

2. Vertical fenestration that receives direct solar radiation for less than 250 hours per year because of shading by permanent external buildings, existing permanent infrastructure, or topography.

8.7.2 Materials

Reported emissions or VOC contents specified below shall be from a representative product sample and conducted with each product reformulation or at a minimum of every 3 years. Products certified under third-party certification programs as meeting the specific emission or VOC content requirements listed below are exempted from this 3-year testing requirement but shall meet all the other requirements as listed below.

8.7.2.1 Adhesives and Sealants

Products in this category include carpet, resilient and wood flooring adhesives, base cove adhesives, ceramic tile adhesives, drywall and panel adhesives, aerosol adhesives, adhesive primers, acoustical sealants, fire stop sealants, HVAC air duct sealants, sealant primers, and caulks. All adhesives and sealants used on the interior of the building (defined as inside of the weatherproofing system and applied on-site) shall comply with the requirements of Section 8.7.2.1.

8.7.2.1.1 Emissions Requirements

Emissions shall be determined according to CA/DHS/EHLB/R-174 (commonly referred to as California Section 01350) and shall comply with the limit requirements for either office or classroom spaces regardless of the space type.

8.7.2.2 VOC Content Requirements

VOC content shall comply with and be determined according to the following limit requirements:

1. Adhesives, sealants, and sealant primers: SCAQMD Rule 1168. HVAC duct sealants shall be classified as "other" category within the SCAQMD Rule 1168 sealants table.

2. Aerosol adhesives: Green Seal Standard GS-36.

Exceptions to Section 8.7.2 include the following:

1. The following solvent welding and sealant products are not required to meet the emissions or the VOC content requirements listed

above: cleaners, solvent cements, and primers used with plastic piping and conduit in plumbing, fire suppression, and electrical systems.

2. HVAC air duct sealants when the air temperature of the space in which they are applied is less than 4.5°C.

8.7.3 Paints and Coatings

Products in this category include sealers, stains, clear wood finishes, floor sealers and coatings, waterproofing sealers, primers, flat paints and coatings, nonflat paints and coatings, and rust-preventive coatings. Paints and coatings used on the interior of the building (defined as inside of the weatherproofing system and applied on-site) shall comply with either Section 4.2.2.1 or Section 4.2.2.2.

8.7.4 Floor Covering Materials

Floor covering materials installed in the building interior shall comply with the following:

1. Carpet: Carpet shall be tested in accordance with and shown to be compliant with the requirements of CA/DHS/EHLB/R-174 (commonly referred to as California Section 01350). Products that have been verified and labeled to be in compliance with Section 9 of the CA/DHS/EHLB/R-174 comply with this requirement.
2. Hard surface flooring in office spaces and classrooms: Materials shall be tested in accordance with and shown to be compliant with the requirements of CA/DHS/EHLB/R-174 (commonly referred to as California Section 01350). Products that have been verified and labeled to be in compliance with SCS-EC10.2 comply with this requirement.

8.7.5 Composite Wood, Wood Structural Panel, and Agrifiber Products

Composite wood, wood structural panel, and agrifiber products used on the interior of the building (defined as inside of the weatherproofing system) shall contain no added urea-formaldehyde resins. Laminating adhesives used to fabricate on-site and shop-applied composite wood and agrifiber assemblies shall contain no added urea-formaldehyde resins. Composite wood and agrifiber products are defined as particleboard, medium-density fiberboard (MDF), wheat board, strawboard, panel substrates, and door cores. Furniture, fixtures, and equipment (FF&E) are not considered base building elements and are not included in this requirement. Emissions for products covered by this section shall be determined according to and shall comply with one of the following:

1. Third-party certification shall be submitted indicating compliance with the California Air Resource Board's (CARB) regulation entitled

Airborne Toxic Control Measure to Reduce Formaldehyde Emissions from Composite Wood Products. A third-party certifier shall be approved by CARB.

2. CA/DHS/EHLB/R-174 (commonly referred to as California Section 01350) and shall comply with the limit requirements for either office or classroom spaces regardless of the space type.

An exception to Section 8.7.5 is structural panel components such as plywood, particleboard, wafer board, and oriented strand board identified as EXPOSURE 1, EXTERIOR, or HUD-APPROVED are considered acceptable for interior use.

8.7.6 Office Furniture Systems and Seating

All office furniture systems and seating installed prior to occupancy shall be tested according to ANSI/BIFMA Standard M7.1 testing protocol and shall not exceed the limit requirements listed in Appendix E of this standard.

8.7.7 Ceiling and Wall Systems

These systems include ceiling and wall insulation, acoustical ceiling panels, tackable wall panels, gypsum wall board and panels, and wall coverings. Emissions for these products shall be determined according to CA/DHS/EHLB/R-174 (commonly referred to as California Section 01350) and shall comply with the limit requirements for either office or classroom spaces regardless of the space type.

8.8 Performance Option

8.8.1 Daylighting Simulation

8.8.1.1 Usable Illuminance in Office Spaces and Classrooms

The design for the building project shall demonstrate an illuminance of at least 30 fc (300 lux) on a plane 1 m above the floor, within 75% of the area of the daylight zones. The simulation shall be made at noon on the equinox using an accurate physical or computer daylighting model.

1. Computer models shall be built using daylight simulation software based on the ray tracing or radiosity methodology.
2. Simulation is to be done using either the CIE overcast sky model or the CIE clear sky model.

An exception to Section 8.8.1.1 is where the simulation demonstrates that existing adjacent structures preclude meeting the illuminance requirements.

8.8.2 Direct Sun Limitation on Work Plane Surface in Offices

It shall be demonstrated that direct sun does not strike anywhere on a work surface of the work plane in any daylit space for more than 20% of the occupied hours during an equinox day in regularly occupied office spaces. If the work surface height is not defined, a height of 0.75 m above the floor shall be used.

8.8.3 Materials

The emissions of all the materials listed below and used within the building (defined as inside of the weatherproofing system and applied on-site) shall be modeled for individual VOC concentrations. The sum of each individual VOC concentration from the materials listed below shall be shown to be in compliance with the limits as listed in Section 8.8.3 of the CA/DHS/EHLB/R-174 (commonly referred to as California Section 01350) and shall be compared to 100% of its corresponding listed limit. In addition, the modeling for the building shall include at a minimum the criteria listed in Appendix F. Emissions of materials used for modeling VOC concentrations shall be obtained in accordance with the testing procedures of CA/DHS/EHLB/R-174 unless otherwise noted below.

1. Tile, strip, panel, and plank products, including vinyl composition tile, resilient floor tile, linoleum tile, wood floor strips, parquet flooring, laminated flooring, and modular carpet tile.
2. Sheet and roll goods, including broadloom carpet, sheet vinyl, sheet linoleum, carpet cushion, wall covering, and other fabric.
3. Rigid panel products, including gypsum board, other wall paneling, insulation board, oriented strand board, medium-density fiberboard, wood structural panel, acoustical ceiling tiles, and particleboard.
4. Insulation products.
5. Containerized products, including adhesives, sealants, paints, other coatings, primers, and other "wet" products.
6. Cabinets, shelves, and work surfaces that are permanently attached to the building before occupancy. Emissions of these items shall be obtained in accordance with ANSI/BIFMA Standard M7.1.
7. Office furniture systems and seating installed prior to initial occupancy. Emissions of these items shall be obtained in accordance with ANSI/BIFMA Standard M7.1.

An exception to Section 8.8.3 is salvaged materials that have not been refurbished or refinished within 1 year prior to installation.

References

1. ASHRAE Guideline 1.1. *HVAC&R technical requirements for the commissioning process*. Latest edition.
2. ASHRAE Standard 52.2—. *Method of testing general ventilation air-cleaning devices for removal efficiency by particle size*. Latest edition.
3. ASHRAE Standard 55—. *Thermal environmental conditions for human occupancy*. Latest edition.
4. ASHRAE Standard 62.1—. *Ventilation for acceptable indoor air quality—Sets the minimum acceptable ventilation requirements*. Latest edition.
5. ASHRAE Standard 90.1—. *Energy standard for buildings except low-rise residential buildings*.
6. Department of Defense Air Force Engineering. 2004. *Design criteria for prevention of mold in Air Force facilities*. Technical Letter ETL 04-3.
7. U.S. General Services Administration. 2010. *P100 facilities standards for the Public Buildings Service*.

9

Energy Efficiency in Air-Conditioned Buildings

To design an optimum HVAC airside system that provides comfort and air quality in the air-conditioned spaces with efficient energy consumption is a great challenge. Air conditioning identifies the conditioning of air for maintaining specific conditions of temperature, humidity, and dust level inside an enclosed space. The conditions to be maintained are dictated by the need for which the conditioned space is intended and comfort of users. So, the air conditioning embraces more than cooling or heating. The comfort air conditioning is defined as "the process of treating air to control simultaneously its temperature, humidity, cleanliness, and distribution to meet the comfort requirements of the occupants of the conditioned space."[1] Air conditioning therefore includes the entire heat exchange operation as well as the regulation of velocity, thermal radiation, and quality of air, as well as the removal of foreign particles and vapors.[2]

Achieving occupant comfort and health is the result of a collaborative effort of environmental conditions, such as indoor air temperature, relative humidity, airflow velocity, pressure relationship, air movement efficiency, contaminant concentration, illumination, sound and noise, and other factors.

Proper understanding of these factors and their respective effects on human comfort and health leads to development of a proper HVAC airside design. Those parameters were previously investigated to define the acceptable design and operating ranges to obtain the comfort and hygiene in the air-conditioned spaces. So, this chapter evaluates the above parameters and the recent progress of HVAC airside design for air-conditioned spaces. The present study aims to define the current status, future requirements, and expectations.

9.1 Air-Conditioned Applications

Indeed, humans spend a great part of their lives in enveloped spaces, which can be artificially conditioned. The air-conditioned applications vary according to the functionality and sensitivity degree of the application. These applications can be divided into residential, commercial, and healthcare applications. The healthcare applications and some sort of the commercial applications have a critical influence on human health. The following sections evaluate the environmental conditions in these applications and review the methods of control and monitoring.

9.1.1 Comfort Levels

9.1.1.1 Preamble

Proper comfort level can be achieved by reaching the optimum conditions of the indoor air temperature, relative humidity, and airflow velocity. Indoor air temperature is one of the most important conditions to provide optimum comfort. The temperature regulatory center in the brain is about 36.8°C at rest in comfort and increases to about 37.4°C when walking and 37.9°C when jogging. An internal temperature less than about 27.8°C can lead to serious cardiac arrhythmia and death, and temperatures greater than 46.1°C can cause irreversible brain damage. Therefore, the careful regulation of body temperature is critical to comfort and health.[1] This condition can inhibit or promote the growth of bacteria and activate or deactivate viruses, in health-care facilities. Some codes and guidelines specify the temperature (only) as a measure of comfort and healthy. Local temperature distributions greatly affect occupant comfort and perception of the environment. Furthermore, high temperatures may cause increased outgassing of toxins from furnishings, finishes, building materials, etc. Alternatively, ambient temperatures that are too cool can cause occupant discomfort such as shivering, inattentiveness, and muscular and joint tension. Relative humidity affects the comfort feeling directly or indirectly by its influence on the temperature. Improper relative humidity conditions may cause a thermal sensation, skin moisture, discomfort, and tactile sensation of fabrics, health, and perception of air quality.

Elevated humidity levels are known to reduce comfort. At lower levels of humidity, thermal sensation is a good indicator of overall thermal comfort and acceptability. Most guidelines specify the range 35 to 50% as the optimum conditions for relative humidity. A few codes raise the upper limit to 60% as the accepted range, but this is not recommended according to the practice. The airflow velocity plays an important role in the comfort sensation and also in the scavenging of the hazards and airborne particles. According to the results of research and the standards specifications, the optimum airflow velocity falls in the range of 0.2 to 0.25 m/s in the occupied zone.

9.1.1.2 Problem Identification

Many of the HVAC applications suffer from poor distribution of the indoor air temperature and relative humidity, as well as incorrect airflow velocities. This poor distribution arises from poor airflow distribution and the presence of thermal drift due to the buoyancy effect.

9.1.1.3 Status Quo

At present, most research recommends experimental and numerical simulations as the perfect tools to obtain the optimum design. The optimization procedure of the HVAC airside design depends on the predictions of the air temperature distribution based on the simulation of different parametric designs using experimentally verified numerical tools. The influence of various ventilation strategies and the vapor generation rate on the characteristics of temperature and moisture distribution is investigated.[3] The significance of the ventilation strategies on the temperature and moisture distributions was found even with the same ventilation rate. Also found was the importance of the airside design and room furnishing in order to ensure a comfortable environment, especially in the displacement ventilation configuration.[4]

The airflow velocity influence was investigated, and it was found that the velocity could be accepted to be as high as 0.35 m/s in the occupied zone,[5] especially in the mixing ventilation, because this value will cause 20% of dissatisfaction among occupancy only. Still, there is no specific method or formula that can be followed to assess the comfort level in the ventilated and air-conditioned spaces. Also, comparisons of comfort between the two different ventilation situations are not general and depend on the configuration of the spaces and the HVAC airside designs.[5] Obviously, the airside design and the configuration of the conditioned space affect the comfort level as well as the space applications, such as the healthcare facilities, which are so complex.[6] The correct specifications of outdoor ambient conditions affect thermal loads, as shown in Figure 9.1, and consequently the comfort level.[7] Some research recommends changing the focus on the effect of the building envelope to reduce the thermal load and enhance thermal comfort.[8]

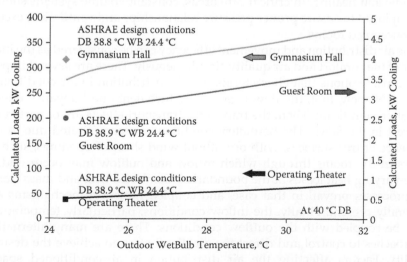

FIGURE 9.1
Outdoor conditions and thermal loads.

9.1.1.4 Conclusions

The comfort conditions depend on many factors beyond the indoor air temperature, relative humidity, and airflow velocity. Comfort conditions depend also on the air distribution pattern and the air movement, but the effect of these factors can be considered close to the air quality more than the comfort level. Indeed, the comfort criteria affect the air quality and the energy conservation in the ventilated and conditioned spaces, as shown in the following sections. The relation between the comfort and air quality is an interchange or mutual relationship.

9.1.2 Air Quality

9.1.2.1 Preamble

Most of the guidelines consider air quality to be the result of a collaborative effort of the environmental conditions presented in the introduction to this chapter. Indeed, in the present literature the air quality is specified by the result of a collaborative effort of the pressure relationship, air movement efficiency, and contaminant concentration. These conditions play an important role to achieve the optimum air quality. Simply, design of a ventilation system must, as much as possible, provide air movement from the clean to the less clean areas. This rule requires great care to design the airside system and select the design of the airside system of the neighborhoods. There are relative interactions between the conditioned neighboring spaces. This criterion is very critical in critical spaces, such as hospitals, and affects the comfort, asepsis, and odor control, and then directly affects the patient hygiene and healing. In critical care areas, constant volume systems should be employed to ensure proper pressure relationships and ventilation, except in unoccupied rooms.

The air distribution and movement efficiency can be considered as the indicator of the comfort and air quality simultaneously. There are several important considerations that characterize the air distribution in air-conditioned spaces. Namely, first, the flow is generally turbulent and buoyancy effects are often significant. Then, the transverse transport effects are of particular interest in the flows. The boundary conditions are complicated due to the presence of free surfaces, with or without wind shear of openings, such as doors in the rooms through which inflow and outflow may occur and of time-varying heat losses at the boundaries. Combined heat and mass transfer processes prevail in that case, and coupled transport mechanisms are generally present. Finally, the inflow conditions, particularly temperature, may be coupled with the outflow conditions. There are many alternative approaches to control and direct the air distribution to achieve the desired quality. Factors affecting the air distribution in air-conditioned spaces should be analyzed to give a better understanding of the nature of the process. In critical applications, such as hospital facilities, air movement takes an extra important role in the controlling of healthy criteria. Undesirable air-flow between rooms and floors is often difficult to control because of open doors, movement of staff and patients, temperature differentials, and stack effect. While some of these factors are beyond practical control, the effect of others may be minimized by terminating shaft openings in enclosed rooms and by designing and balancing air systems to create positive or negative air pressure within certain rooms and certain areas.[9]

Contaminants can be classified in four broad headings, each of which represents a wide variety of pollutants: organic and inorganic compounds, particulate matter, and biological contaminants.

It should be understood that these classifications are intended to facilitate the categorization of contaminants. Although the pollutants are classified into these categories, certain contaminants may belong to two or more classifications, depending upon their nature.[10] The classification of organic compounds represents chemical compounds that contain carbon-hydrogen bonds in their basic molecular structure. Their sources can be either natural products or synthetics, especially those derived from oil, gas, and coal. Organic contaminants may exist in the form of gas (vapor), liquid, or as solid particles in the atmosphere, food, or water.[11] Inorganic compounds are those that do not contain carbon-hydrogen bonds in their molecular structure. They include carbon dioxide, sulfur dioxide, nitrogen oxides, carbon monoxide, ozone, lead, sand, metal, ammonia, and some particulate matter. A complex mixture of organic and inorganic substances formulates particulate matter, each with diverse physical and chemical properties. Furthermore, they represent a wide variety of substances, ranging in size from 0.005 to 100 mm in aerodynamic diameter, including asbestos, dust, mold, pollen, and dander.

The danger of particulate matter is its ability to become contaminated by other ambient sources, increasing health risks to individuals who are exposed to respirable suspended particles (RSPs). Particles falling into this category are usually less than 10 mm in aerodynamic diameter. As mentioned previously, particles smaller than 5 mm are capable of bypassing the respiratory defenses. Biological contaminants are generally referred to as microbes or microorganisms. Biological contaminants are minute particles of living matter produced from a variety of sources. For the most part, sources of biological contaminants are found outdoors; however, many may occur in both outdoor and indoor environments. The variety of biological compounds that may be present in the ambient environment is immense. Therefore, exposure to an increased concentration creates a potential health risk to susceptible individuals.

At extremes of the exposure range for light, heat, cold, and sound, organ dysfunction is measurable, and disease, frostbite, burns, and noise-induced hearing loss occur. Some transitions between healthy and disease states are more difficult to delineate. Pain from bright light, erythema from heat, and nausea from vibration represent reversible effects but are interpreted by health professionals as abnormal. Air quality must also be maintained to provide a healthy, comfortable indoor environment. Sources of pollution exist in both the internal and external environments. The air quality is controlled by removal of the contaminant or by dilution.[2] ASHRAE Standard 1981[12] prescribes both necessary quantities of ventilation for various types of occupancies and methods of determining the proportions of outside air and recirculated air. If the level of contaminants in outdoor air exceeds that for minimum air quality standards, extraordinary measures must be used. The ASHRAE standard provides the necessary recommendations for the residential, commercial, and industrial applications. Although proper air conditioning designs are helpful in the prevention and treatment of diseases, the application of air conditioning to health facilities presents many specific problems. Those are not encountered in the conventional comfort conditioning design.

9.1.2.2 Problem Identification

The contaminant concentration mainly depends on two factors: air pressure relationship and air movement efficiency. So the optimum design of these two factors leads to accepted concentration and safe distribution of the contaminant. Actually, most of guidelines known to date don't restrict any airside design for each application. This gives a large tolerance and many design alternatives, which are not totally perfect.

9.1.2.3 Status Quo

The comfort and air quality are investigated with the aid of experimental and numerical techniques, to represent the relationship between the thermal conditions and air quality.[13] It was found that thermal conditions affect the air quality, and therefore any recommended numerical models should account for balanced thermal conditions. This would affect the discrepancies between measured and simulated results, and consequently create a more generalized numerical formula of the air characteristics.

The effect of thermal loads and cooling strategies on the airflow pattern in an office was investigated by applying mixing ventilation,[14] as well as those results indicating the effect of supply conditions on the airflow pattern.[6] The ventilation system performance was found to have high importance in energy savings and clean environment. The effect of the heat and contaminant source location on the ventilation performance was also considered. It was found that the optimum ventilation performance is achieved when these sources are located near the exhaust opening.[15] The air quality is generally influenced by airborne and contaminant generation in healthcare applications, especially in critical sites such as isolation and surgery rooms. Several researches investigated the airborne particle control in operating rooms using the numerical techniques.[16–19] These researches indicated that the particle distribution depends on the particle source location, airside design, and furniture and equipment distribution. The airflow turbulence also has strong influence on the contaminant concentration.

9.1.2.4 Conclusions

Air movement efficiency is mainly based on two factors: the air pressure relationship with the other neighborhood spaces and the airside design.

Differential air pressure can be maintained only in an entirely sealed room. Therefore, it is important to obtain a reasonably close fit of all doors and seal all walls and floor penetrations between pressurized areas. This is best accomplished by using weather stripping and drop bottoms on doors. The opening of a door between two areas instantaneously reduces any existing pressure differential between them to such a degree that its effectiveness is nullified. When such openings occur, a natural interchange of air takes place between the two rooms due to turbulence created by the door opening and closing combined with personal ingress/egress. For critical areas requiring both maintenance of pressure differentials to adjacent spaces and personnel movement between the critical area and adjacent spaces, the use of appropriate air locks or anterooms is indicated. In general, outlets supplying air to sensitive ultraclean areas should be located on the ceiling, and perimeter or several exhaust outlets should be near the floor. This arrangement provides a downward movement of clean air through the breathing and working zones to the floor area for exhaust. Infectious isolation rooms should have supply air above and near the doorway and exhaust air from near the floor, behind the patient's bed. This arrangement is such that clean air first flows to parts of the room where workers or visitors are likely to be, and then flows across the infectious source and into the exhaust. Thus, noninfected persons are not positioned between the infectious source and the exhaust location. The bottom of the return or exhaust openings should be at least 75 mm above the floor.

9.2 Energy Efficiency

9.2.1 Preamble

Energy crisis in the early 1970s forced the development of energy-conserving strategies in a variety of industries. Sustainability and energy efficiency continue to be strong issues in this time of limited resources. Therefore, the implementation of energy-conserving strategies in HVAC systems must be balanced with occupant comfort and health. Few guidelines gave specific recommendations about energy saving in HVAC systems, but these recommendations don't meet all requirements and design varieties. Indeed, in hot and humid climates the outdoor conditions play an important role in energy consumption. Also, the utilization strategies of the conditioned air in the conditioned space play an important role in saving energy consumption.

9.2.2 Problem Identification

Until now, the guidelines and design standards have not provided restricted utilization strategies of the conditioned air in spaces. Indeed, this situation creates several inefficient systems, and consequently expensive energy invoices. In some critical facilities, such as hospitals, HVAC designers face the problem of balancing between healthy conditions and energy utilization.

9.2.3 Status Quo

The relationship between HVAC system designs and optimum conditions and optimum energy utilization is still under investigation. In recent research,[20,21] the effect of ventilation design on comfort and energy utilization has been investigated. The effect of displacement ventilation on the humidity gradient in a factory located in a hot and humid region is illustrated.[20] A strong dependence relationship between the correct supplying conditions and comfort was found. Indeed the displacement ventilation is recommended as an energy-efficient system; the resulting relative humidity and temperature gradients give designers suitable tolerances to select more economic supply conditions.[20] In recent years, new design traditions of the ventilation systems, such as under-floor ventilation systems, are growing to overcome the problems of the current systems. Under-floor air supply is recommended as an alternative to ceiling air supply in office buildings to overcome the lack of flexibility in the ceiling systems and improve the comfort conditions.[21] It is noticeable that those who advised the under-floor system recommend it due to its capability to reduce energy consumption due to the operational characteristics of supplied air.

9.2.4 Conclusions

As the optimization of energy consumption is a new trend, the achievement of this level needs a new investigation trend in the scientific research. Actually, energy utilization mainly depends on the optimum utilization of the conditioned air in conditioned spaces.

9.3 Evaluation Indices

9.3.1 Preamble

The evaluation indices of the comfort, air quality, and energy utilization efficiency can be divided to two main categories: empirical indices based on the experimental techniques and numerical indices based on the numerical techniques. The most common indices provide the required evaluation of the air characteristics at individual positions (or in other scope, at individual points) in the indoor environment.

9.3.2 Problem Identification

Until now, the evaluation of comfort, air quality, and energy utilization efficiency was performed only at individual positions (local evaluation). Still, there is no general global evaluation index for several characteristics, such as the airflow movement and the contaminant concentration and its influence on the occupancy health. Actually, the airflow distribution pattern currently plays the role of global evaluation index. On the other hand, there is no global evaluation index capable of evaluating comfort, air quality, and energy utilization efficiency simultaneously. Actually, this global index will aid the HVAC designers to achieve the optimum design according to the optimum indoor air quality levels.

9.3.3 Status Quo

The experimental techniques play an important role to yield a complete view of air characteristic patterns in air-conditioned spaces. The numerical techniques play an important role in dealing with the parametric designs. This is a very powerful tool to predict the indoor air quality of any space with the complex configurations and sensitive functionality. But in contrast, this technique is not capable of introducing a global evaluation of the conditioned space by itself.[8,22–28] There is a new trend in recent years that is based on the integration of artificial intelligent tools with numerical tools to replace the human being in decision making. The genetic algorithm is one of the artificial intelligent techniques that is usually integrated with the numerical tool, and is found to be highly suitable for exploring alternatives from different areas of the design spaces to enhance and evaluate the HVAC performance.[29] Indeed, this attempt does not create a flexible evolution index. The analytical and numerical techniques are integrated in many researches[30] to enhance the analytical technique performance to predict the ventilation performance. Indeed, the analytical method needs massive effort to obtain the results. In some situations, the performance of the analytical method is less than the expected. The neuro-fuzzy technique is also integrated with the numerical method[17] to evaluate the ventilation performance, depending on the conclusion, which is "the evaluation action is a mental process."

9.3.4 Conclusions

Present evaluation indices do not introduce the complete evaluation for the simulated ventilation design or for the air-conditioned spaces. The individual evaluation indices can perform the duty partially, stalling the progress in this field. Indeed, the future of these indexes is guided by the capabilities of the hardware and software.

9.4 Observations

9.4.1 Technical Observations

- In general, the best investigation technique of the status in the air-conditioned spaces is that based on experimental and numerical investigations. Currently, the ultimate goal of experimental investigations is to create a validated numerical tool for the prediction of the characteristics of the indoor environment (Figure 9.2). So experimental work plays a remarkable role in obtaining the airflow characteristics of the ventilation and air conditioning systems. The disadvantages are the huge amount of time required and the expensive efforts of this technique.

- There are many trials to create global or general quantitative indices to evaluate the comfort, air quality, and energy utilization efficiency resulting from the airside design parameters changing with no significant progress. This arises from the complex nature of the indoor environment, the complex nature of each target (comfort, air quality, energy utilization efficiency) in the global index, and the conflict among them in the many situations.

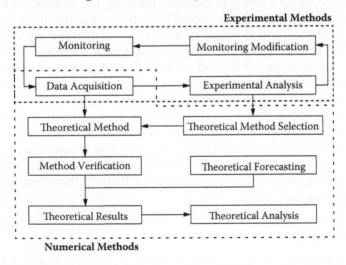

FIGURE 9.2
Experimental and numerical techniques.

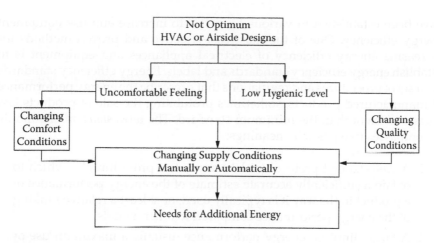

FIGURE 9.3
Causes of energy waste.

9.4.2 Environmental Observations

- Actually, the dependence on supply conditions only to save energy is not valuable. This trend leads researchers to unrealistic results. One should investigate the roots of the energy-wasting problem in air-conditioned spaces due to HVAC systems (Figure 9.3). Indeed, the optimum airside configuration design can save energy directly.

- Airflow characteristics, especially the humidity ratio and the contaminant concentration, are the main comfort and air quality factors, respectively, in global indoor air quality.

- The airflow distribution pattern needs new development to increase the performance of ventilation systems. For example, as of today, there is no final solution for dealing with recirculation zones and dead zones in air-conditioned spaces, but only some recommendations about the airside designs.

- Actually, the new traditions in the HVAC and airside designs are required to enable the designers to select the optimum design from a variety. But this new trend will add new responsibilities to assess these new designs and their suitability with the required internal conditions.

9.5 International Energy Standards

9.5.1 General

As the world becomes increasingly dependent on electrical appliances and equipment, energy consumption rapidly rises every year. Many programs have been established in various countries to increase end use equipment energy efficiency. One of the most cost-effective and proven methods for increasing energy efficiency of electrical appliances and equipment is to establish energy efficiency standards and labels. Energy efficiency standards are a set of procedures and regulations that prescribe the energy performance of manufactured products, sometimes prohibiting the sale of products less energy efficient than the minimum standard. The term *standard* commonly encompasses two possible meanings:

1. A well-defined protocol (or laboratory test procedure) by which to obtain a sufficiently accurate estimate of the energy performance of a product in the way it is typically used, or at least a relative ranking of the energy performance compared to other models
2. A target limit on energy performance (usually a maximum use or minimum efficiency) formally established by a government-based agency upon a specified test standard

Energy efficiency labels are informative labels affixed to manufactured products indicating a product's energy performance (usually in the form of energy use, efficiency, or cost) in order to provide consumers with the data necessary for making informed purchases. Energy labels serve as a complement to energy standards. They provide consumers information that allows those who care to select more efficient models. Labels also allow utility companies and government energy conservation agencies to offer incentives to consumers to buy the most energy-efficient products. The effectiveness of energy labels is highly dependent on how information is presented to the consumer.

9.5.2 Rationale and Benefits

Energy efficiency in developed and developing countries plays an important role in achieving global sustainable development. Energy efficiency improvements can slow the growth in energy consumption, save consumers money, and reduce capital expenses for energy infrastructure. Energy consumption is growing rapidly in these countries, yet energy efficiency remains far below levels in developed countries. Energy efficiency improvements can slow the growth in energy consumption, save consumers money, and reduce capital expenses for energy infrastructure.[31-40] For most developing countries, the foreign exchange needed to finance energy sector expansion is a significant drain on reserves. Additionally, energy efficiency reduces local environmental impacts, such as water and air pollution from power plants, and mitigates greenhouse gas emissions. Standards and labeling programs provide enormous energy savings potential that can direct developing countries toward sustainable energy use. Improved end use efficiency from standards and labeling programs can contribute significantly to developing economies. The main benefits are as follows:

1. Less need to build new power plants. The cost of saving 1 kWh of energy through energy efficiency programs has proven much less expensive than producing 1 kWh of energy by building a new power plant.

2. Reduced greenhouse gas emissions. Less energy production means less carbon dioxide emissions from power plants. This contributes to environmental benefits such as slowing down environmental pollution and global warming and preserving natural resources and the ecosystem.

3. Improved competitiveness for local manufacturers. Local companies that upgrade the efficiency of their products can compete better with multinational companies, especially with lower production costs.

4. Higher consumer disposable income. Less spending on electric bills increases consumer purchasing power for other products, which helps local businesses.

5. Increased cash flow in the local economy. With higher disposable income, consumers are more willing to spend, thus injecting money into the local economy.

6. Improved trade balance. Decrease in energy demand will reduce the consumption of indigenous resources (i.e., natural gas and oil), allowing more to be exported (for Lebanon, less to be imported). Increased export earnings (or less import spending) help alleviate the trade deficit of Arabian countries.

7. Avoided future energy deficit as power demand rises. Energy exporting countries have become net importers due to dramatic increases in electricity demand. Energy efficiency programs can help slow down the demand and prevent an energy deficit in the future.

9.5.3 International Standards for Energy-Efficient Buildings

International standards and labeling programs are defined as a set of elements that ensure that energy efficiency standards and labeling efforts are effective, appropriate, strengthened over time, and sustained. The building blocks fall into two categories: technical/policy and process.

9.5.3.1 Technical/Policy

1. Accredited testing facilities. Facilities should be internationally accredited, staffed with competent testing personnel, and have the capacity to test models in a timely manner.

2. Appropriate testing procedures. Testing procedures are the methods by which the energy efficiency level of a product is deduced. The selected procedures should reasonably reflect the usage patterns and climate particular to a country. This builds consumer confidence that test results accurately reflect the energy usage they will experience.

3. Energy labels. Standards and labels can be established separately or as complementary programs. Many types of labeling programs exist.

4. Energy efficiency standards. Standards can be mandatory or voluntary and based on either maximum energy consumption or minimum energy efficiency.

5. Energy policy framework. An energy policy framework that is conducive to energy efficiency is critical to the longevity of a national standards program. Supportive policies include government procurement requirements, voluntary programs, incentives to manufacturers, consumer awareness campaigns, and demand-side management and integrated resource planning.

9.5.3.2 Proposed Process

1. Compliance with voluntary and mandatory standards and labeling requirements must be ensured through a credible enforcement scheme to guarantee program effectiveness. Program evaluation will inform necessary program modifications, justify further activities, and provide the documentation necessary to sustain the standards and labeling programs over the long term.

2. The legislative process should ensure that standards and labels are periodically reviewed and raised (ratcheted upward) as the overall product efficiency on the market improves. The changes will mostly depend on the results of program evaluation.

3. In the program design and improvement process, input from all stakeholders (government, private companies, consumer associations, etc.) should be considered. Cooperation between the stakeholders is the key to the success of programs. However, the local and national governments must also hold their decision final, after carefully considering all suggestions.

The following steps are required for the energy-efficient design of a built environment:

1. Develop methodologies to measure and access energy efficiency in buildings.

2. Develop methodologies for energy declaration of the buildings.

3. Develop reference values (key numbers) or systems for benchmarking.

4. Provide a labeling system for selected buildings.

9.5.4 Efforts by the International Organization for Standardization (ISO)

Currently two of the technical committees, ISO TC 163 and 205, are focusing on achieving the above targets in a built environment; these can be detailed as follows.

9.5.4.1 Technical Committee ISO TC 163

This committee is for standardization in the field of building and civil engineering works:

- Thermal and hygrothermal performance of materials, products, components, elements, and systems, including complete buildings
- Thermal insulation materials, products, and systems for building and industrial application, including insulation of installed equipment in buildings, covering and including:

1. Test and calculation methods for heat and moisture transfer, temperature, and moisture conditions
2. Test and calculation methods for energy use in buildings
3. Test and calculation methods for heating and cooling loads in buildings
4. In situ test methods for thermal, hygrothermal, and energy performance of buildings
5. Building components
6. Input data for calculations, including climatic data
7. Specifications for thermal insulation materials, products, and systems with related tests
8. Methods and conformity criteria
9. Terminology
10. General review and coordination of work on thermal and hygrothermal performance within ISO

9.5.4.2 Technical Committee ISO TC 205

This committee is for standardization in the design of new buildings and retrofit of existing buildings for acceptable indoor environment and practicable energy conservation and efficiency. Indoor environment includes air quality and thermal, acoustic, and visual factors.

Excluded are the following:

- Other ergonomic factors
- Methods of measurement of air pollutants and thermal, acoustic, and lighting properties
- Methods of testing for performance and rating of building environmental equipment and thermal insulation

9.6 Energy-Efficient Buildings

9.6.1 Energy Declaration of Buildings

The primary use of energy declarations is to

1. Create consciousness of energy efficiency in buildings and also improve the knowledge of energy use in buildings
2. Use the information to determine if the building works as well as possible with regard to its technical design
3. Use the information for benchmarking
4. Use the information for suggesting measures and recommendations for reducing the energy use
5. Provide the information necessary to make calculations of the environmental impact due to the energy use, e.g., CO_2 emission
6. Describe selected energy properties of the building
7. Give the basis for a common energy performance certification of a building

Depending on the purpose of the energy declaration, different procedures can be of interest. Different actors need different information; e.g., for benchmarking and for explaining the CO_2 emission, it can be enough to read the total energy supply to the building and only adjust these figures to normal outdoor climate and the heated area. For giving relevant advice to the property owner regarding which measures are cost-effective, a very careful examination and calculation of the building's energy balance is necessary.

One way to proceed is to make the energy calculation in different steps for existing buildings. The first is to collect measured energy use, e.g., from energy bills, and make a benchmark to decide if the actual building is better or worse than similar buildings. If the energy use seems to be higher than the average for a comparable group of buildings, a second step is to make a careful energy calculation that can be compared to the measured energy use. This has to be done for identifying what kind of measures can be recommended in order to reduce the energy use in the building. For benchmarking, it can perhaps also be of interest to compare the measured energy use in the building examined with the estimated energy use in a building that is built with the best available technology. Alternatively, compare with a building that meets the requirement in the existing building codes. Some important aspects necessary to take into consideration when developing a common tool for energy declaration of buildings are discussed. The discussion here focuses on residential buildings, but similar principles are relevant for other types of buildings.

9.6.2 Energy Declaration of Existing Buildings

For most existing buildings the energy use usually is well known via the energy bills. The construction details, on the other hand, are not very well documented. Calculations of the energy use will thus in most cases be very uncertain, and difficulties will occur when giving relevant recommendations of cost-effective energy conservation measures.

The energy declaration needs a combination of tools for calculations based on the information from the energy bills. For existing buildings the energy declarations can be based on both measurements and calculations in order to fulfill the purposes mentioned above. Normally the heating bills are based on measured heat delivered to the building. In most cases, the energy for domestic hot water is included in the heating bill. To get the total energy use in a building, the electricity to run the building and household electricity have to be added. In electrically heated houses, common in, e.g., Sweden, the house owner receives just one bill covering the total energy use. To make the measured values objective and comparable, several corrections and calculations are necessary:

- It is necessary to check that the indoor thermal comfort and air quality meet agreed requirements.
- Corrections of the heat use to normal outdoor climate—primarily outdoor temperature (maybe also solar heat gain).
- Necessary corrections for internal heat gains, e.g., differences in household electricity.

9.6.3 Energy Declaration of New Buildings

The energy declaration of a new building has to be based on calculations. In comparison with existing buildings, the knowledge of the building construction is very good. The most critical part for the outcome of calculations is the choice of input data. To achieve good comparability, a common procedure to determine input data like energy for domestic hot water, household electricity or electricity for the activity in the building, choice of indoor temperature, electricity for operating the building, energy for lighting, etc., has to be developed.

In many countries many new buildings have very low energy demand for heating. The solar heat gain, internal gains and energy losses from equipment, etc., cover the major part of the heating demand. In several buildings the internal gains, etc., are so large that cooling is necessary even in temperate climates. For buildings with large glassed areas or large internal gains the energy calculations need to be done on an hourly basis. Many modern apartment blocks, offices, education buildings, and restaurants need very little heat supply from the heating system and very often need air conditioning to get an acceptable thermal comfort.

9.6.4 Issues for International Collaboration

- Develop standardized tools for the calculation of the energy performance of buildings, taking into account ISO 13790, which covers many aspects but is not yet complete.
- Define system boundaries for the different building categories and different heating systems.
- Prepare models for expressing requirements on indoor air quality, thermal comfort in winter and, when appropriate, summer, visual comfort, etc.
- Develop transparent systems to determine necessary input data for the calculations, including default values on internal gains.
- Provide transparent information regarding output data (reference values, benchmarks, etc.).
- Define comparable energy-related key values (kWh/m^2, $kWh/person$, $kWh/apartment$, $kWh/produced unit$, etc.). The areas/volumes need to be defined.
- Develop a method to translate net energy used in the building to primary energy and CO_2 emissions.
- Develop a common procedure for an energy performance certificate.
- Develop and compile relevant standards applicable for each individual building category.

9.7 Concluding Remarks

It can be concluded that it is important to incorporate an energy performance directive as a standard in our region. Such a goal will aid energy savings in large buildings and set regulations to energy-efficient designs that are based on standard calculation methods. It is recommended to

1. Develop standardized tools for the calculation of the energy performance of buildings

2. Define system boundaries for the different building categories and different heating systems

3. Prepare models for expressing requirements on indoor air quality, thermal comfort in winter and, when appropriate, summer, visual comfort, etc.

4. Define comparable energy-related key values (kWh/m^2, kWh/person, kWh/apartment, kWh/produced unit, etc.) and develop a common procedure for an energy performance certificate

5. Design, construct, and operate a solar decathlon (building) that can meet the rural and desert requirements and save the diminishing fossil fuel sources

---

References

1. ASHRAE Handbook. 2001. *Fundamentals*. ASHRAE, Atlanta, GA.
2. Stoecker, W. F., and Jones, J. W. 1985. *Refrigeration and air conditioning*, second edition. TATA McGraw-Hill Publishing Company LTD, New York.
3. Liu, J., Aizawa, Y., and Yoshino, H. 2002. Experimental and numerical study on simultaneous temperature and humidity distributions. ROOMVENT 2002, 169–172.
4. Naydenov, K., Pitchurov, G., Langkilde, G., and Melikov, A. K. 2002. Performance of displacement ventilation in practice. ROOMVENT 2002, 483–486.
5. Jacobsen, T. S., Hansen, R., Mathiesen, E., Nielsen, P. V., and Topp, C. 2002. Design method and evaluation of thermal comfort for mixing and displacement ventilation. ROOMVENT 2002, 209–212.
6. Kameel, R., and Khalil, E. E. 2001. Numerical computations of the fluid flow and heat transfer in air-conditioned spaces, NHTC2001-20084, 35th National Heat Transfer Conference, Anaheim, California.
7. Khalil, E. E. 2000. Computer aided design for comfort in healthy air conditioned spaces. *Proceedings of Healthy Buildings 2000, 2*, 461–466.
8. Nakamura, Y., and Fujikawa, A. 2002. Evaluation of thermal comfort and energy conservation of an ecological village office. ROOMVENT 2002, 413–416.
9. ASHRAE. 1999. *Applications*. ASHRAE, Atlanta, GA.
10. Lee, T. G., De Biasio, D., and Santini, A. 1996. Health and the built environment: Indoor air quality. Vital Signs Curriculum Materials Project, Health and the Built Environment. The University of Calgary, Calgary, Alberta.
11. Rea, W. J. 1992. *Chemical sensitivity*, vol 1. Boca Raton: Lewis Publishers.
12. ASHRAE standards 55-1981. ASHRAE, Atlanta, GA.
13. Holmberg, S., and Einberg, G. 2002. Flow behaviour in a ventilated room—measurements and simulations. ROOMVENT 2002, 197–200.

14. Corgnati, S. P., Fracastoro, G. V., and Perino, M. 2002. Influence of cooling strategies on the air flow pattern in an office with mixing ventilation. ROOMVENT 2002, 165–168.
15. Cho, Y., and Awbi, H. B. 2002. Effect of heat source location in a room on the ventilation performance. ROOMVENT 2002, 445–448.
16. Liu, Y., and Moser, A. 2002. Airborne particle concentration control for an operating room. ROOMVENT 2002, 229–232.
17. Kameel, R. 2002. Computer aided design of flow regimes in air-conditioned operating theatres. PhD thesis, Cairo University, Cairo, Egypt.
18. Kameel, R., and Khalil, E. E. 2002. Prediction of flow, turbulence, heat-transfer and air humidity patterns in operating theatres. ROOMVENT 2002, 69–72.
19. Kameel, R., and Khalil, E. E. 2002. Prediction of turbulence behaviour using k-· model in operating theatres. ROOMVENT 2002, 73–76.
20. Kosonen, R. 2002. Displacement ventilation for room air moisture control in hot and humid climate. ROOMVENT 2002, 241–244.
21. Leite, B. C. C., and Tribess, A. 2002. Analysis of under floor air distribution system: Thermal comfort and energy consumption. ROOMVENT 2002, 245–248.
22. Collignan, B., Couturier, S., and Akoua, A. A. 2002. Evaluation of ventilation system efficiency using CFD analysis. ROOMVENT 2002, 77–80.
23. Somarathne, S., Kolokotroni, M., and Seymour, M. 2002. A single tool to assess the heat and airflows within an enclosure: Preliminary test. ROOMVENT 2002, 85–88.
24. Tibaut, P., and Wiesler, B. 2002. Thermal comfort assessment of indoor environments by means of CFD. ROOMVENT 2002, 97–100.
25. Braconnier, R., Fontaine, J. R., and Bonthoux, F. 2002. Use of predictive ventilation to evaluate the emission rates of pollutant sources in an enclosure and to reconstruct the associated concentration field. ROOMVENT 2002, 133–136.
26. Sorenson, D. N., and Weschler, C. J. 2002. Modeling chemical reactions in the indoor environment by CFD. ROOMVENT 2002, 149–152.
27. Kuwabara, K., Mochida, T., Nagano, K., and Shimakura, K. 2002. Evaluation of thermal sensation in urban environment. ROOMVENT 2002, 417–420.
28. Willem, H. C., and Sekhar, S. C. 2002. Correlation of indoor air quality measurements and CFD simulations—findings from a case study in the tropics. ROOMVENT 2002, 421–424.
29. Choudhary, R., and Malkawi, A. 2002. Integration of CFD and genetic algorithms. ROOMVENT 2002, 121–124.
30. Ross, A. D. 1999. On the effectiveness of ventilation. PhD thesis, Eindhoven University of Technology, Eindhoven, Netherlands.
31. Khalil, E. E. 2005. Energy performance of buildings directive in Egypt: A new direction. *HBRC Journal*, 1.
32. Medhat, A. A., and Khalil, E. E. 2006. Thermal comfort meets human acclimatization in Egypt. *Proceedings of Healthy Buildings*, 2, 25.
33. Khalil, E. E. 2006. Energy performance of commercial buildings in Egypt: A new direction. *Proceedings of Energy 2030*, Abu Dhabi, UAE.
34. Kosonen, R. 2002. Displacement ventilation for room air moisture control in hot and humid climate. ROOMVENT 2002, 241–244.
35. Leite, B. C. C., and Tribess, A. 2002. Analysis of under floor air distribution system: Thermal comfort and energy consumption. ROOMVENT 2002, 245–248.
36. Kameel, R., and Khalil, E. E. 2002. Prediction of turbulence behaviour using k-ε model in operating theatres. ROOMVENT 2002, 73–76.

37. Khalil, E. E. 2008. Arab-air conditioning and refrigeration code for energy-efficient buildings. *Arab Construction World*, 28, 8, 24–26.
38. Khalil, E. E. 2008. Air conditioning and refrigeration code for energy-efficient buildings in the Arab world. *Journal of Kuwait Society of Engineers*, 100, 94–95.
39. ISO publications. January 2009.
40. Federal Register. Rules and regulations 72565, vol. 72, no. 245. December 21, 2007.

Appendix A: Nomenclature

Symbols, Terms, Units, and Subscripts

Symbol	Description	Unit
A	Surface area	m²
A_c	Cross-sectional area of a duct	m²
ADPI	Air distribution performance index	Dimensionless
BF	Bypass factor	Dimensionless
C	Unit thermal conductance	W/(m²/°C)
C	Loss coefficient	Dimensionless
C_d	Discharge coefficient	Dimensionless
C_p	Pressure coefficient	Dimensionless
C_v	Flow coefficient	Dimensionless
c	Specific heat	J/kg °C
c	Sound velocity in the air	m/s
c_p	Specific heat at constant pressure	J/kg °C
CLF	Cooling load factor	Dimensionless
CLTD	Cooling load temperature difference	°C
COP	Coefficient of performance	Dimensionless
cfm	Volume flow rate	ft³/min
D	Diameter	m
DD	Degree-days	°C day
DR	Daily range of temperature	°C
EER	Energy efficiency ratio	Dimensionless
f	Friction factor	Dimensionless
G	Irradiation	W/m²
g	Local acceleration due to gravity	m/s²
H	Head	m
H	Enthalpy	J
h	Heat transfer coefficient	W/m²°C
h	Specific enthalpy	J/kg
h_{fg}	Latent heat of vaporization or condensation	J/kg
hp	Horsepower	hp
k	Thermal conductivity	Wm/m²°C
L	Total length	m
L_p	Sound pressure level	dB
L_W	Sound power level	dB
L_{WA}	A-weighted sound power level	dB(A)
LH	Latent heat	W
LMTD	Log mean temperature difference	°C

Continued

Symbols, Terms, Units, and Subscripts (*Continued*)

Symbol	Description	Unit
M	Molecular weight	kg/kmol
m	Mass	kg
$\dot{m}$	Mass flow rate	kg/s
NC	Noise criteria	Dimensionless
NTU	Number of transfer units	Dimensionless
Nu	Nusselt number	Dimensionless
P	Circumference length	m
P_E	Electrical motor input power	W
p	Static pressure	Pa
p_d	Dynamic pressure	Pa
p_i	Partial pressure of component i	Pa
Pr	Prandtl number	Dimensionless
Q	Total energy input	J
$\dot{Q}$	Heat transfer rate	W
q	Heat transfer per unit mass	J/kg
R	Gas constant	J/(kg K)
r	Thermal resistance	m² °C/W
r	Radius	m
r_f	Fouling resistance	m² °C/W
$\bar{R}$	Universal gas constant = 8.314	J/mol.K
Re	Reynolds number	Dimensionless
rpm	Revolutions per minute	Dimensionless
SC	Shading coefficient	Dimensionless
SH	Sensible heat	W
SHF	Sensible heat factor	Dimensionless
T	Temperature	°C
T	Absolute temperature	K
TH	Total heat	W
U	Internal energy	J
U	Overall heat transfer coefficient	W/m² °C
u	Velocity	m/s
V	Volume	m³
$\dot{V}$	Volume flow rate	m³/s
v	Specific volume	m³/kg
W	Work	J
$\dot{W}$	Power	W
ω	Moisture content, also known as humidity ratio	kg$_v$/kg$_a$
x	Length	m
y	Length	m
z	Length	m
Δp	Pressure drop	Pa

Symbols, Terms, Units, and Subscripts (*Continued*)

Symbol	Description	Unit
η_h	Humidification efficiency	Dimensionless
μ	Dynamic viscosity	kg/ms
ν	Kinematic viscosity	m²/s
r	Density	kg/m³
φ	Relative humidity	%

Subscripts

a	Air, dry air, added
adp	Apparatus dew point
as	Adiabatic saturation
atm	Atmospheric
avg	Average
b	Branch
c	Cool, coil, cold, condenser, cross section or minimum free area, ceiling
cl	Center line
cw	Chilled water
cwr	Chilled water return
cws	Chilled water supply
D	Diameter
d	Design
db	Dry bulb
dp	Dew point
e	Evaporator
f	Friction, fouling
fg	Phase change (liquid/vapor)
fl	Floor
g	Saturated vapor
h	Hot
i	At state i when i is a letter or number, component i in a gas mixture
m	Mixed
o	Outdoor
p	Constant pressure
r	Rejected
R	Room or recirculated
s	Saturation, supply
t	Total
v	Water vapor, constant volume
vs	Saturated water vapor at a given dry-bulb temperature
w	Water
wb	Wet bulb

Continued

Symbols, Terms, Units, and Subscripts (*Continued*)

Symbol	Description	Unit
Superscripts		
*	Properties of saturated air at a given wet-bulb temperature	
Abbreviations		
AC	Air conditioning	
AHU	Air handling unit	
ASHRAE	American Society of Heating, Refrigerating, and Air-Conditioning Engineers	
BHP	Brake horsepower	
DX	Direct expansion, dry expansion	
FCU	Fan coil unit	
HVAC&R	Heating, ventilating, air conditioning, and refrigeration	
SI	International System of Units	

Appendix B: Climatic Conditions in Luxor

TABLE A2.1

Luxor Design Conditions

Month	dbt	wbt	dp	RH	epsi	V	h
January, max	23	16.79	13.11	54	9.7	0.87	47.65
February, max	26	17.77	13.06	45	9.67	0.88	50.63
March, max	30	19.14	13.22	36	9.77	0.89	54.95
April, max	35	21.91	15.75	32	11.54	0.91	64.59
May, max	39	22.79	15.3	25	11.2	0.92	67.81
June, max	41	24.15	16.98	25	12.49	0.93	73.16
July, max	41	25.09	18.79	28	14.02	0.93	77.11
August, max	41	25.69	19.9	30	15.05	0.93	79.75
September, max	39	25.37	20.21	34	15.34	0.92	78.44
October, max	35	22.89	17.61	36	13.01	0.91	68.36
November, max	30	21.74	18.04	49	13.37	0.9	64.16
December, max	25	18.63	15.24	55	11.16	0.88	53.42

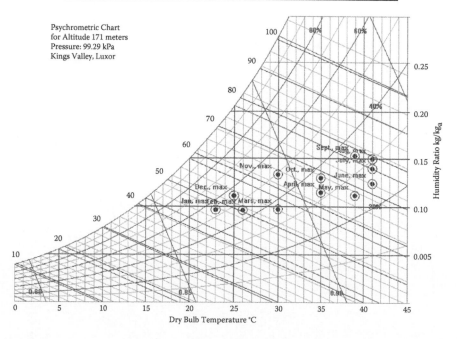

Psychrometric Chart
for Altitude 171 meters
Pressure: 99.29 kPa
Kings Valley, Luxor

Appendix B: Climatic Conditions in Luxor

TABLE A2.1

Luxor Design Conditions

Month		db	wb	rh	RH	apm	V	u

Appendix C: Glossary

Access hatch: An access hatch is defined as a door, thereby allowing it to meet less stringent envelope requirements. If not defined as a door, it would need to be insulated as a roof or wall, depending on where it was located.

Accessible (as applied to equipment): Admitting close approach; not guarded by locked doors, elevations, or other effective means. *See also* readily accessible.

Adjusted lighting power: Lighting power, ascribed to a luminaire(s) that has been reduced by deducting a lighting power control credit based on use of an automatic control device(s).

Adopting authority: Agency or agent that adopts this standard.

Air-conditioned floor area: Area equipped with air conditioning equipment measured at floor level from the interior surfaces of the walls.

Air-conditioned space: Space equipped with air conditioning equipment.

Air economizer: Duct and damper arrangement and automatic control system that together allow a cooling system to supply outside air to reduce or eliminate the need for mechanical cooling during mild or cold weather.

Air handling unit: An encased assembly consisting of sections containing a fan or fans and other necessary equipment to perform one or more of the following functions: circulating, filtration, heating, cooling, heat recovery, humidifying, dehumidifying, and mixing of air.

Alteration: Rearrangement, replacement, or addition to a building or its systems and equipment; routine maintenance and service or a change in the building's category shall not constitute an alteration.

Area factor: Multiplying factor that adjusts the unit power density for spaces of various sizes to account for the impact of room configuration on lighting power utilization.

Area of the space (A): Horizontal lighted floor area of a given space measured from the inside of the perimeter walls or partitions, at the height of the working surface.

Automatic: Self-acting, operating by its own mechanism when actuated by some impersonal intervention, such as a change in current strength, pressure, temperature, or mechanical configuration.

Automatic control device: Device capable of automatically turning loads off and on without manual intervention.

Average daily temperature: The average of the 24 h readings of temperatures.

Ballast: Device used in conjunction with an electric discharge lamp to cause the lamp to start and operate under the proper circuit conditions of voltage, current, wave form, electrode heat, etc.

Ballast efficacy factor: Ratio of relative light output to the power input.

Ballast efficacy factor—fluorescent: Ratio of the ballast factor expressed as a percent to the power input in watts, at specified test conditions.

Ballast, electronic: Ballast constructed using electronic circuitry.

Ballast factor (BF): Ratio of commercial ballast lamp lumens to reference ballast lamp lumens, used to correct the lamp lumen output from rated to actual; ratio of the lumen output of a lamp-ballast combination to the lumen output of the same lamp in combination with a piece of laboratory equipment called a reference reactor. Because the ballast may be designed to operate more than one lamp type, the same ballast model may have more than one ballast factor value.

Ballast, hybrid: Ballast constructed using a combination of magnetic core and insulated wire winding and electronic circuitry.

Ballast, magnetic: Ballast constructed with magnetic core and a winding of insulated wire.

Boiler: Device to raise the temperature of a fluid or generate steam.

Boiler capacity: Rated heat output of the boiler, at the design inlet and outlet conditions and rated fuel or energy input.

Budget building design: Computer representation of a hypothetical design based on the actual proposed building design. This representation is used as the basis for calculating the energy cost budget.

Building: Construction as a whole, including its envelope and all technical building systems.

Building area: Greatest horizontal area of a building above grade within the outside surface of exterior walls or within the outside surface of exterior wall and the centerline of fire walls.

to the building by its users and occupants.

Building energy cost: Computed annual energy cost of all purchased energy for the building.

Building entrance: Any doorway set of doors, turnstiles, or other form of portal that is ordinarily used to gain access.

Building envelope: Elements of a building that enclose conditioned spaces through which thermal energy may be transferred to or from the exterior or to or from unconditioned spaces.

Building exit: Any doorway set of doors, or other form of portal that is ordinarily used for emergency egress or convenience exit.

Building grounds lighting: Lighting provided through a building's electrical service for parking lot, site, and roadway, pedestrian pathway, loading dock, exterior architectural lighting, and security applications.

Building official: The official authorized to act on behalf of the authority having jurisdiction.

Building type: The classification of a building by usage.

Coefficient of performance (COP), cooling mode: Ratio of the rate of heat removal to the rate of energy input in consistent units, for a complete cooling system or factory assembled equipment, as tested under a nationally recognized standard or designated operating conditions.

Coefficient of performance (COP), heat pump—heating mode: Ratio of the rate of heat delivered to the rate of energy input, in consistent units, for a complete heat pump system under designated operating conditions. Supplemental heat shall not be considered when checking compliance with the heat pump equipment COPs.

Coefficient of utilization (CU): Ratio of lumens from a luminaire calculated as received on the work plane to the lumens emitted by the luminaire lamps alone factored by room surface reflectances and room dimensions. *See also* room cavity ratio.

Compact fluorescent lamp: Fluorescent lamp of a small compact shape, with a single base that provides the entire mechanical support function.

Conditioned space: Enclosure served by an air distribution system.

Continuous insulation: Insulation that is continuous across all structural members without any thermal bridges, excluding fasteners and service openings. It is installed on the interior, exterior, or integral to any opaque surface of the building envelope.

Control: To regulate the operation of equipment.

Control device: Specialized device used to regulate the operation of equipment.

Control loop, local: Control system consisting of a sensor, controller, and controlled device.

Control point: Quantity of equivalent ON or OFF switches ascribed to a device used for controlling the light output of a luminaire(s) or lamp(s).

Cooldown: Reduction of space temperature down to occupied set point after a period of shutdown or setup.

Cooling: Removal of latent or sensible heat.

Cooling degree-day: *See* degree-day.

Cooling design temperature: Outdoor dry-bulb temperature for sizing cooling systems, equal to the temperature that is exceeded 2.5% of the number of hours during the nominal cooling season (June through September in the northern hemisphere) in a typical weather year.

Cooling design wet-bulb temperature: Outdoor wet-bulb temperature for sizing cooling systems and evaporative heat rejection systems, such as cooling towers.

Daylight sensing control (DS): Device that automatically regulates the power input to electric lighting near the fenestration to maintain the desired workplace illumination, thus taking advantage of direct or indirect sunlight.

Daylit area: Area under horizontal fenestration (skylight) or adjacent to vertical fenestration (window) described as follows:

Daylit area, horizontal: Area under horizontal fenestration (skylight) with a horizontal dimension in each direction equal to the skylight dimension in that direction plus either the floor-to-ceiling height, the distance to the nearest 1 m or higher opaque partition, or one-half the distance to an adjacent skylight or vertical glazing clerestory, whichever is least.

Daylit area, vertical: Area adjacent to vertical fenestration (window) with one horizontal dimension that extends into the space either at a distance of 4.5 m or to the nearest 1 m or higher opaque partition, whichever is less; another horizontal dimension equal to the width of the window plus either 0.6 m on each side, the distance to an opaque partition, or one-half the distance to an adjacent skylight or window, whichever is least.

Daylit space: Space bounded by vertical planes rising from the boundaries of the daylit area on the floor to the above floor or roof.

Daylit zone: Types of daylit zones are as follows:

> **Under skylights:** Area under each skylight whose horizontal dimension in each direction is equal to the skylight dimension in that direction plus either the floor-to-ceiling height, the dimension to an opaque partition, or one-half the distance to an adjacent skylight or vertical glazing, whichever is least.

> **At vertical glazing:** Area adjacent to vertical glazing that receives daylighting from the glazing. For purposes of this definition and unless more detailed daylighting analysis is provided, the daylighting zone depth is assumed to extend into the space a distance of 4.5 m or to the nearest opaque partition, whichever is less. The daylighting zone width is assumed to be the width of the window plus either 0.6 m on each side, the distance to an opaque partition, or one-half the distance to an adjacent skylight or vertical glazing, whichever is least.

Dead band (dead zone): Range of values within which an input variable can be varied without initiating any noticeable change in the output variable.

Decorative lighting: *See* lighting, decorative.

Degree-day: Difference in temperature between the outdoor mean temperature over a 24 h period and a given base temperature. For the purposes of determining building envelope requirements, the classifications are defined as follows:

> **Degree-day, cooling:** For any 1 day, when the mean temperature is more than the base temperature, there are as many degree-days as degrees Celsius temperature difference between the mean temperature for the day and the base temperature. Annual cooling degree-days (CDDs) are the sum of the degree-days over a calendar year.

> **Degree-day, heating:** For any 1 day, when the mean temperature is less than the base temperature, there are as many degree-days as degrees Celsius temperature difference between the mean temperature for the day and base temperature. Annual heating degree-days (HDDs) are the sum of the degree-days over a calendar year.

Dehumidification: Controlled reduction of water vapor from the air.

Demand energy: Energy to be delivered to provide the required service with an ideal system to the end user.

Design capacity: Output capacity of a system or piece of equipment at design conditions.

Design conditions: Specified indoor environmental conditions, such as temperature, relative humidity, lighting level, etc., required to be produced and maintained by a system and under which the system must operate.

Design energy consumption: Estimated annual energy usage of a proposed building design.

Design energy costs: Estimated annual energy expenditure of proposed building design.

Direct digital control (DDC): Type of control where controlled and monitored analogue or binary data (e.g., temperature, contact closures) are converted to digital format for manipulation and calculations by a digital computer or microprocessor, and then converted back to analogue or binary form to control physical devices.

Distribution system: Conveying means, such as ducts, pipes, and wires, to bring substances or energy from a source to the point of use. The distribution system includes such auxiliary equipment as fans, pumps, and transformers.

Efficiency: Performance at specified rating conditions.

Efficiency, HVAC system: Ratio of the useful energy output (at the point of use) to the energy input in consistent units for a designated time period, expressed in percent.

Emittance: Ratio of the radiant heat flux emitted by a specimen to that emitted by a blackbody at the same temperature and under the same conditions.

Enclosed space: Volume substantially surrounded by solid surfaces, such as walls, floors, roofs, and openable devices, such as doors and operable windows.

Energy: Capability for doing work; having several forms that may be transformed from one to another, such as thermal (heat), mechanical (work), electrical, or chemical.

Energy carrier: Substance of phenomenon that can be used to produce mechanical work or heat or to operate chemical or physical processes.

Energy cost: Cost of energy by unit and type of energy as proposed to be supplied to the building at the site, including variations such as time of day, seasonal, and rate of usage.

Energy cost budget: Maximum allowable estimated annual energy expenditure for a proposed building.

Energy efficiency ratio (EER): Ratio of net equipment (cooling or heating) capacity to total rate of electric input under designated operating conditions. When consistent units are used, this ratio becomes equal to COP. *See also* coefficient of performance.

Energy efficiency ratio for buildings (EERB): Ratio of energy required (ER) and energy used (EU).

Energy factor for water heater (EF): Measure of water heater overall efficiency.

Energy management system: Control system designed to monitor the environment and the use of energy in a facility and to adjust the parameters of local control loops to conserve energy while maintaining a suitable environment.

Energy need for heating or cooling: Heat to be delivered to or extracted from a conditioned space by a heating or cooling system to maintain the intended temperature during a given period of time.

Energy need for domestic hot water: Heat to be delivered to the domestic water to raise its temperature from the cold network temperature to the prefixed delivery temperature at the delivery point.

Energy performance of a building: Calculated or measured amount of energy actually used or estimated to meet the different needs associated with a standard use of the building, which may include internal energy use for heating, cooling, ventilation, domestic hot water, and lighting.

Envelope performance factor: Trade-off value for the building envelope performance compliance option calculated using the procedures specified in the system's performance trade-off. For the purposes of determining building envelope requirements, the classifications are defined as follows:

Base envelope performance factor: Building envelope performance factor for the base design.

Proposed envelope performance factor: Building envelope performance factor for the proposed design.

Energy resource: Energy taken from a source that is depleted by extraction (e.g., fossil fuels) and is required to achieve the building performance and comfort over a given period of time, including HVAC, lighting, occupancy, domestic hot water, etc.).

Energy use for space heating or cooling: Energy input to the heating or cooling system to satisfy the energy need for heating or cooling, respectively.

Enthalpy: Thermodynamic property of a substance defined as the sum of its thermodynamic energy plus the quantity PV, where P is the pressure and V is its volume; formerly called total heat and heat content.

Equipment: Devices for comfort conditioning, electric power, lighting, transportation, or service water heating, including, but not limited to, furnaces, boilers, air conditioners, heat pumps, chillers, water heaters, lamps, luminaries, ballasts, elevators, escalators, or other devices or installations.

Exfiltration: Uncontrolled passage of air from a space through leakage paths in the shell of that space.

Exterior envelope: *See* building envelope.

Exterior lighting power allowance: Calculated maximum lighting power allowance for an exterior area of a building or facility, in watts.

Exterior sheltered building envelope: The elements of a building that separate conditioned spaces from the exterior.

Facade area, vertical: Area of the facade, including one horizontal roof area, overhangs, and cornices, measured in elevation in a vertical plane parallel to the plane of the face of the building.

Feeder conductors: Wires that connect the service equipment to the branch circuit breaker panels.

Fenestration: Any light-transmitting section in a building wall or roof. The fenestration includes glazing material (which may be glass or plastic), framing (mullions, muntins, and dividers), external shading devices, internal shading devices, and integral (between glass) shading devices.

Fixture: Component of a luminaire that houses the lamp(s), positions the lamp, shields it from view, and distributes the light. The fixture also provides for connection to the power supply, which may require the use of a ballast.

Floor area, gross: Floor area of heated or cooled spaces excluding nonhabitable cellars and unheated spaces, including the floor area on all stories if more than one.

Fluorescent lamp: Low-pressure electric discharge lamp in which a phosphor coating transforms some of the ultraviolet energy generated by the discharge into light.

General lighting: *See* lighting, general.

General service lamp: Class of incandescent lamps that provide light in virtually all directions. General service lamps are typically characterized by bulb shape, such as A, standard; S, straight side; F, flame; G, globe; and PS, pear straight.

Glazed wall system: Category of site-assembled fenestration products that includes, but is not limited to, curtain walls and solariums.

Gross building envelope floor area: Gross floor area of the building envelope, but excluding slab-on-grade floors.

Gross conditioned floor area: Gross floor area of conditioned spaces.

Gross exterior wall area: Gross area of exterior walls separating a conditioned space from the outdoors or from unconditioned spaces as measured on the exterior above grade. It consists of the opaque wall (excluding vents and grills), including between floor spandrels, peripheral edges of flooring, window areas including sash, and door areas.

Gross floor area over outside or unconditioned spaces: Gross area of a floor assembly separating a conditioned space from the outdoors or from unconditioned spaces as measured from the exterior faces of exterior walls or from the centerline of walls separating buildings. The floor assembly shall be considered to include all floor components through which heat may flow between indoor and outdoor or unconditioned environments.

Gross lighted floor area: Gross floor area of lighted spaces.

Gross lighted area: Sum of the total lighted areas of a building measured from the inside of the perimeter walls for each floor of the building.

Gross roof area: Gross area of a roof assembly separating a conditioned space from the outdoors or from unconditioned spaces, measured from the exterior faces of exterior walls or from the centerline of walls separating buildings. The roof assembly shall be considered to include all roof or ceiling components though which heat may flow between indoor and outdoor environments, including skylights but excluding service openings.

Gutter: Space available for wiring inside panel boards and other electric panels. A separate wire way used to supplement wiring spaces in electric panels.

Heat: Form of energy that is transferred by virtue of a temperature difference or a change in state of a material.

Heated space: *See* space.

Heating degree-day: *See* degree-day.

Heating design temperature: Outdoor dry-bulb temperature for sizing heating systems.

Heating seasonal performance factor: Total heating output of a heat pump during its normal annual usage period for heating, divided by the total electric energy input during the same period.

Heating, ventilating, air-conditioning system: Equipment, distribution systems, and terminals that provide, either collectively or individually, the processes of heating, ventilating, or air conditioning to a building or portion of a building.

High-intensity discharge lamp: Electric discharge lamp in which light is produced when an electric arc is discharged through a vaporized metal such as mercury or sodium. Some HID lamps may also have a phosphor coating that contributes to the light produced or enhances the light color.

Historic building: Building or space that has been specifically designated as historically significant by the adopting authority.

Humidistat: Automatic control device used to maintain humidity at a fixed or adjustable set point.

HVAC system efficiency: *See* efficiency, HVAC system.

Incandescent lamp: Lamp in which light is produced by a filament heated to incandescence by an electric current.

Indirectly conditioned space: Enclosed space within the building that is not a heated or cooled space, whose area-weighted heat transfer coefficient to heated or cooled spaces exceeds that to the outdoors or to unconditioned spaces, or through which air from heated or cooled spaces is transferred at a rate exceeding three air changes per hour. *See also* heated space, cooled space, and unconditioned space.

Infiltration: Uncontrolled inward air leakage through cracks and crevices in any building element and around windows and doors of a building caused by pressure differences across these elements due to factors such as wind, inside and outside temperature differences (stack effect), and imbalance between supply and exhaust air systems.

Insolation: Rate of solar energy incident on a unit area with a given orientation.

Installed interior lighting power: Power in watts of all permanently installed general, task, and furniture lighting systems and luminaries as indicated on plans and specifications.

Integrated part-load value: Single-number figure of merit based on part-load energy efficiency ratio (EER); coefficient of performance (COP) expressing part-load efficiency for air conditioning and heat pump equipment on the basis of weighted operation at various load capacities for the equipment.

Interior lighting power allowance: *See* lighting power allowance.

Interior unit lighting power allowance—prescriptive: Allotted interior lighting power for each individual building type, in W/m^2.

Interior unit lighting power allowance—system performance: Allotted interior lighting power for each individual space, area, or activity in a building, in W/m^2.

Isolation devices: Devices that isolate HVAC zones so that they can be operated independently of one another. Isolation devices include, but are not limited to, separate systems, isolation dampers, and controls providing shutoff at terminal boxes.

Lamp: Generic term for a man-made light source often called a bulb or tube.

Lamp-ballast efficacy: Lumens produced by a lamp-ballast combination (the product of rated lamp lumen output and the relative light output of the lamp-ballast combination) divided by the watts of input power, expressed in lumens per watt.

Lamp efficacy: Quotient of the total light (lumens) emitted and the total lamp power input (watts), expressed in lumens per watt.

Lamp lumens, rated: Light output of a lamp as published in manufacturer's literature.

Lamp wattage, rated: Power consumption of a lamp as published in manufacturer's literature.

Lighting: Lighting that provides a substantially uniform level of illumination throughout an area.

Lighting, decorative: Lighting that is purely ornamental and installed for aesthetic effect. Decorative lighting shall not include general lighting.

Lighting efficacy: Quotient of the total lumens emitted from a lamp or lamp-ballast combination divided by the watts of input power, expressed in lumens per watt.

Lighting, general: General lighting shall not include decorative lighting or lighting that provides a dissimilar level of illumination to serve a specialized application or feature within such area.

Lighting power allowance: Lighting allowance that includes the following:

  Exterior lighting power allowance: Maximum lighting power in watts allowed for the exterior of a building.

  Interior lighting power allowance: Maximum lighting power in watts allowed for the interior of a building.

Lighting power budget: The lighting power, in watts, allowed for an interior or exterior area or activity.

Lighting power control credit: Credit applied to that part of the connected lighting power of a space that is turned off or dimmed by automatic control devices. It gives the specific value of lighting watts to subtract from the connected interior lighting power when establishing compliance with the interior lighting power allowance.

Lighting power density: Maximum lighting power per unit area of a building classification of space function.

Lighting system: Group of luminaries circuited or controlled to perform a specific function.

Lumen: Radiometrically, it is determined from the radiant power. Photometrically, it is the luminous flux emitted within a unit solid angle (one steradian) by a point source having a uniform luminous intensity of one candela.

Lumen maintenance control: Device that senses the illumination level and causes an increase or decrease of illuminance to maintain a preset illumination level.

Luminaire: Complete lighting unit consisting of a lamp(s) together with the housing designed to distribute the light, position and protect the lamps, and connect the lamps to the power supply.

Luminance: Density of the luminous flux incident on a surface. It is the quotient of a luminous flux by the area of the surface when the latter is uniformly illuminated.

Manual: Requiring personal intervention for control.

Marked (nameplate) rating: Design load operating conditions of a device as shown by the manufacturer on the nameplate or otherwise marked on the device.

Mean daily temperature: One-half the sums of the minimum daily temperature and maximum daily temperature.

Mechanical heating: Raising the temperature of a gas or liquid by use of fossil fuel burners, electric resistance heaters, heat pumps, or other systems that require energy to operate.

Mechanical refrigeration: Reducing the temperature of a gas or liquid by using vapor compression, absorption, and desiccant dehumidification combined with evaporative cooling, or another driven thermodynamic cycle. Indirect or direct evaporative cooling alone is not considered mechanical refrigeration.

Motor efficiency, minimum: Minimum efficiency occurring in a population of motors of the same manufacturer and rating.

Motor efficiency, nominal: Median efficiency occurring in a population of motors of the same manufacturer and rating.

Motor power, rated: Rated output power from the motor.

Net thermal efficiency, generation: The ratio between the heat or cooling demand of the distribution system and the fuel heat input energy requirements for heating or cooling. Energy to be delivered to the heating or cooling system to satisfy the heat demand of the building.

Occupancy sensor: Device that detects the presence or absence of people within an area and causes lighting, equipment, or appliances to be regulated accordingly.

Optimum start controls: Controls that are designed to automatically adjust the start time of an HVAC system each day with the intention of bringing the space to desired occupied temperature levels immediately before scheduled occupancy.

Orientation: The direction an envelope element faces, i.e., the direction of a vector perpendicular to and pointing away from the surface outside of the element. For vertical fenestration, the two categories are north oriented and all other.

Outdoor (outside) air: Air that is outside the building envelope or is taken from outside the building that has not been previously circulated through the building.

Packaged terminal air conditioner: Factory-selected wall sleeve and separate unencased combination of heating and cooling components, assemblies, or sections. It may include heating capability by hot water, steam, or electricity and is intended for mounting through the wall to serve a single room or zone.

Packaged terminal heat pump: Packaged terminal air conditioner (PTAC) capable of using the refrigeration system in a reverse cycle or heat pump mode to provide heat.

Party wall: Fire wall on an interior lot line used or adapted for joint service between two buildings.

Performance indicator: Performance indicators are calculated for the designed indoor environmental conditions and are generally used to indicate how the building performs from energy, carbon dioxide emissions, and cost standpoints. Systems can be identified as HVAC, domestic hot water (DHW), lighting, automation, and control systems. Different indicators are listed here for illustrations; selections would depend on the aim and objective of the project.

1. Energy demand of the building envelope:
   a. Total energy demand   [kWh]
   b. Energy demand/Floor unit   [kWh/m²]
2. Integrated performance including systems:
   a. Total energy used   [kWh]
   b. Integrated intensity of energy used = 2a/Floor unit [kWh/m²]
   c. Building (active) efficiency = Energy demand/Energy used

   [–]
3. Primary (weighted) energy performance: Same as 2 but multiply energy used from any energy carrier with weighted primary coefficient
4. CO_2 emission:
   a. Total CO_2 emission per year   [g]
   b. Relative CO_2 emission 4b = 4a/Floor unit   [g/m²]
5. Cost efficiency:
   a. Design cost per total energy cost and floor unit
   b. Design cost per floor unit
   c. Design cost per total energy used

Design cost could be defined in different ways:
   - Investment costs
   - Global economics, including annual costs
   - Cost related to design lifetime of the building, including any operational life cycle costs (LCC)

Permanently installed: Equipment that is fixed in place and is not portable or movable.

Piping: System for conveying fluids, including pipes, valves, strainers, and fittings.

Plenum: Enclosure that is part of the air distribution system and is distinguished by having almost uniform air pressure. A plenum often is formed in part or in total by portions of the building.

Power: In connection with machines, it is the time rate of doing work; in connection with the transmission of energy of all types, it is the rate at which energy is transmitted. It is measured in watts (W).

Power adjustment factor: Modifying factor that adjusts the effective connected lighting power (CLP) of a space to account for the use of energy-conserving lighting control devices.

Power factor: Ratio of total real power in watts to the apparent power (root mean square volt amperes).

Prescribed assumption: Fixed value of an input to the standard calculation procedure.

Primary air system: Central air moving heating and cooling equipment that serves multiple zones through mixing boxes, variable air volume (VAV) boxes, or reheats coils.

Primary energy: Energy that has not been subjected to any conversion or transformation process.

Primary energy efficiency: The ratio between the energy required and the primary energy requirements to assume for the energy used.

Process energy: Energy consumed in support of a manufacturing, industrial, or commercial process other than conditioning spaces and maintaining comfort and amenities for the occupants of a building.

Process load: Load on a building resulting from the consumption or release of process energy.

Projection factor: Ratio of the horizontal depth of the external shading projection divided by the sum of the height of the fenestration and the distance from the top of the fenestration to the bottom of the farthest point of the external shading projection, in consistent units. See Figure C.1.

Note: *Projection factor* is a term to indicate how much shading an overhang provides for vertical fenestration. As an example, if a 1.8 m horizontal overhang was placed right at the top of a 3 m store window, the projection factor would be 1.8(3.0 + 0.0) = 0.60. However, if the horizontal overhang was located 0.0 m above the window, then the projection factor would be 1.8(3.0 + 0.6) = 0.50.

Proposed design: Computer representation of the actual proposed building design or portion thereof that incorporates standard requirements. This representation is used as the basis for calculating the design energy cost.

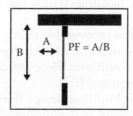

FIGURE C.1
Projection factor.

Prototype building: Generic building design of the same size and occupancy type as the proposed design that complies with the prescriptive requirements of this standard and has prescribed assumptions used to generate the energy budget concerning shape, orientation, HVAC, and other system designs.

Pump system energy demand (pump system power): Sum of the nominal power demand (nameplate horsepower at nominal motor efficiency) of motors of all pumps that are required to operate at design conditions to supply fluid from the heating or cooling source to all heat transfer devices (e.g., coils, heat exchanger) and return it to the source.

Radiant comfort heating: System in which temperatures of room surfaces are adjusted to control the rate of heat loss by radiation from occupants.

Radiant heating system: Heating system that transfers heat to objects and surfaces within the heated space primarily (greater than 50%) by infrared radiation.

Rated lamp lumens: *See* lamp lumens, rated.

Rated lamp wattage: *See* lamp wattage, rated.

Rated motor power: *See* motor power, rated.

Readily accessible: Capable of being reached quickly for operation, renewal, or inspections without requiring those to whom ready access is requisite to climb over or remove obstacles or to resort to portable ladders, chairs, etc. In public facilities, accessibility may be limited to certified personnel through locking covers or by placing equipment in locked rooms.

Recirculating system: Domestic or service hot water distribution system that includes a closed-circulation circuit designed to maintain usage temperatures in hot water pipes near terminal devices (e.g., lavatory faucets, shower heads) in order to reduce the time required to obtain hot water when the terminal device valve is opened. The motive force for circulation is either natural (due to water density variations with temperature) or mechanical (recirculation pump).

Recoiling: Lowering the temperature of air that has been previously heated by a mechanical heating system.

Recoverable energy: Part of the energy losses, from the space and domestic hot water system or lighting, which can be recovered to lower the energy required.

Recovered energy: Part of the recoverable energy. Energy utilized from an energy utilization system that would otherwise be wasted (not contributing to a desired end use). Recovered energy may contribute to reduce the energy required (ER).

Reference building: Specific building design that has the same form, orientation, and basic systems as the proposed design and meets all the criteria of the prescriptive compliance method.

Reflectance: Ratio of the light reflected by a surface to the light incident upon it.

Reflector lamp: Class of incandescent lamps that have an internal reflector to direct the light. Reflector lamps are typically characterized by reflective characteristics such as R, reflector; ER, ellipsoidal reflector; PAR, parabolic aluminized reflector; MR, mirrorized reflector; and others.

Reheating: Raising the temperature of air that has been previously cooled by either mechanical refrigeration or an economizer system.

Reset: Automatic adjustment of the controller set point to a higher or lower value.

Roof area, gross: Area of the roof measured from the exterior faces of walls or from the centerline of party walls. *See* roof and wall.

Room air conditioner: Encased assembly designed as a unit to be mounted in a window or through a wall, or as a console. It is designed primarily to provide direct delivery of conditioned air to an enclosed space, room, or zone. It includes a prime source of refrigeration for cooling and dehumidification and a means for circulating and cleaning air. It may also include a means for ventilating and heating.

Room area: For lighting power determination purpose, the area of a room or space shall be determined from the inside face of the walls or partitions measured at the work plane height.

Room cavity ratio (RCR): Factor that characterizes room configuration as a ratio between the walls and ceiling and is based upon room dimensions.

Sash crack: Sum of all perimeters of all ventilators, sashes, or doors based on overall dimensions of such parts expressed in meters (counting two adjacent lengths of perimeter as one).

Semiexterior sheltered building envelope: The elements of a building that separate conditioned space from unconditioned space (as far as it is not designed for human occupancy) or that enclose semiheated spaces through which thermal energy may be transferred to or from the exterior, or to or from unconditioned spaces, or to or from conditioned spaces.

Sequence: Consecutive series of operations.

Service systems: All energy-using or -distributing components in a building that are operated to support the occupant or process functions housed therein (including HVAC, service water heating, illumination, transportation, cooking or food preparation, laundering, or similar functions).

Service water heating: Heating water for domestic or commercial purposes other than space heating and process requirements.

Service water heating demand: Maximum design rate of water withdrawal from a service water heating system in a designated period of time (usually an hour or a day).

Set point temperature: Internal (minimum) temperature, as fixed by the control system in normal heating mode, or internal (maximum) temperature, as fixed by the control system in normal cooling mode.

Shell building: Building for which the envelope is designed, constructed, or both prior to knowing the occupancy type. *See also* speculative building.

Single-zone system: System that provides heating or cooling to a single space or a group of spaces that have thermal load requirements sufficiently similar that desired conditions can be maintained throughout by a single temperature control device.

Site-recovered energy: Waste energy recovered at the building site that is used to offset consumption of purchased fuel or electrical energy supplies.

Site-solar energy: Thermal, chemical, or electrical energy derived from direct conversion of incident solar radiation at the building site and used to offset consumption of purchased fuel or electrical energy supplies. For the purposes of applying this standard, site-solar energy shall not include passive heat gain through fenestration systems.

Solar energy source: Source of thermal, chemical, or electrical energy derived from direct conversion of incident solar radiation at the building site.

Solar heat gain coefficient (SHGC): Ratio of the solar heat gain entering the space through the fenestration area to the incident solar radiation. Solar heat gain includes directly transmitted solar heat and absorbed solar radiation, which is then reradiated, conducted, or convected into the space. *See* fenestration area.

Space: Enclosed space within a building. The classifications of spaces are as follows for the purpose of determining building envelope requirements:

    Conditioned space: Cooled space, heated space, or indirectly conditioned space, defined as follows.

Standard energy calculation procedure: Energy simulation model and a set of input assumptions that account for the dynamic thermal performance of the building; it produces estimates of annual energy consumption for heating, cooling, ventilation, lighting, and other uses.

Standby loss: Standby loss (in percent per hour) expressed as a ratio of the heat loss per hour to the heat content of the stored water above room temperature. When calculating heat loss, all energy sources required to maintain stored water at a preset temperature are considered, including the primary heating energy (gas, oil, or electricity) and any electricity consumed by a blower motor, controls, circulating pump, etc.

Task conditioning: Air conditioning that provides individual comfort for a specific surface or area.

Task lighting: Lighting that provides illumination for specific visual functions and is directed to a specific surface or area.

Task location: Area of the space where significant visual functions are performed and where lighting is required above and beyond that required for general ambient use.

Technical building systems: Technical equipment for heating, cooling, ventilation, domestic hot water, lighting, and electricity production.

Terminal: Device by which energy from a system is finally delivered, e.g., registers, diffusers, lighting fixtures, faucets, etc.

Terminal element: Device by which the transformed energy from a system is finally delivered, i.e., registers, diffusers, lighting fixtures, faucets, etc.

Thermal block: Collection of one or more HVAC zones grouped together for simulation purposes. Spaces need not be contiguous to be combined within a single thermal block.

Thermal mass: Materials with mass heat capacity and surface area capable of affecting building loads by storing and releasing heat as the interior or exterior temperature and radiant conditions fluctuate. *See also* wall heat capacity.

Thermal zone: Part of the (controlled) space with a given set point temperature, throughout which the internal temperature is assumed to have negligible spatial variation.

Thermostat: Automatic control device used to maintain temperature at a fixed or adjustable set point.

Thermostatic control: Automatic control device or system used to maintain temperature at a fixed or adjustable set point.

Tinted glazing: Coloring that is integral with the glazing material. Tinting does not include surface applied films such as reflective coatings, applied either in the field or during the manufacturing process.

Total lighting power allowance: Calculated lighting power allowed for the interior and exterior space areas of a building or facility.

Unitary cooling equipment: One or more factory-made assemblies that normally include an evaporator or cooling coil and a compressor and condenser combination. Units that perform a heating function are also included.

Unitary heat pump: One or more factory-made assemblies that normally include an indoor conditioning coil, compressor(s), and an outdoor refrigerant-to-air coil or refrigerant-to-water heat exchanger. These units provide both heating and cooling functions.

Unit energy costs: Costs for units of energy or power purchased at the building site. These costs may include energy costs as well as costs for power demand as adopted by the authority having jurisdiction.

Unit lighting power allowance: Allotted lighting power for each individual building type is W/m^2.

Unit power density: Lighting power density, in W/m^2, of an area or activity.

Variable air volume system: Systems that control the dry-bulb temperature within a space by varying the volume of supply air to the space.

Vent damper: Device intended for installation in the venting system of an individual, automatically operated, fossil fuel-fired appliance in the outlet or downstream of the appliance draft control device, which is designed to automatically open the venting system when the appliance is in operation and to automatically close off the venting system when the appliance is in a standby or shutdown condition.

Ventilation: Process of supplying or removing air by natural or mechanical means to or from any space. Such air is not required to have been conditioned. (See Figure 3.4.)

Ventilation air: That portion of supply air that comes from outside (outdoors), plus any recirculated air, to maintain the desired quality of air within a designated space. *See also* outdoor air.

Wall: That portion of the building envelope, including opaque area and fenestration, that is vertical or tilted at an angle of 60° from horizontal or greater. This includes above- and below-grade walls, between floor spandrels, peripheral edges of floors, and foundation walls.

Wall area, gross: Area of the wall measured on the exterior face from the top of the floor to the bottom of the roof.

Warm-up: Increase in space temperature to occupied set point after a period of shutdown or setback.

Water economizer: System by which the supply air of a cooling system is cooled directly or indirectly or both by evaporation of water or other appropriate fluid (in order to reduce or eliminate the need for mechanical refrigeration).

Water heater: Closed vessel in which water is heated and is withdrawn for use external to the system, including the apparatus by which heat is generated and all controls and devices necessary to prevent service water from exceeding safe limits.

Window-to-wall ratio (WWR): Ratio of the fenestration area to the gross exterior wall area.

Zone, HVAC: Space or group of spaces within a building with heating and cooling requirements that are sufficiently similar so that desired conditions (e.g., temperature) can be maintained throughout using a single sensor (e.g., thermostat or temperature sensor).

Index

Printed and bound by CPI Group (UK) Ltd, Croydon, CR0 4YY

18/10/2024

01776262-0004